PRINCIPLES OF BIOCHEMICAL TESTS IN DIAGNOSTIC MICROBIOLOGY

GEORGE WASHINGTON UNIVERSITY
PAUL HIMMELFARB HEALTH SCIENCES LIBRARY
2300 EYE STREET, N. W.
WASHINGTON, D. C. 20037

PRINCIPLES OF BIOCHEMICAL TESTS
IN DIAGNOSTIC MICROBIOLOGY

TECHNIQUES IN
PURE AND APPLIED MICROBIOLOGY

CARL-GÖRAN HÉDEN, Editor

Karolinska Institute, Stockholm

G. Sermonti, *Genetics of Antibiotic-Producing Microorganisms*

V. B. D. Skerman, Editor, *Abstracts of Microbiological Methods*

Noel R. Rose and Pierluigi E. Bigazzi, Editors, *Methods in Immunodiagnosis*

Carl-Göran Hedén and Tibor Illéni, Editors, *Automation in Microbiology and Immunology*

Carl-Göran Hedén and Tibor Illéni, Editors, *New Approaches to the Identification of Microorganisms*

Brij M. Mitruka, *Gas Chromatographic Applications in Microbiology and Medicine*

Donna J. Blazevic and Grace Mary Ederer, *Principles of Biochemical Tests in Diagnostic Microbiology*

PRINCIPLES OF BIOCHEMICAL TESTS IN DIAGNOSTIC MICROBIOLOGY

DONNA J. BLAZEVIC, M.P.H.

Professor

GRACE MARY EDERER, M.P.H.

Associate Professor

Department of Laboratory Medicine and Pathology
University of Minnesota

PRINCIPLES OF
BIOCHEMICAL TESTS
IN DIAGNOSTIC
MICROBIOLOGY

A WILEY BIOMEDICAL PUBLICATION

JOHN WILEY & Sons
New York • London • Sydney • Toronto

Library of Congress Cataloging in Publication Data:

Blazevic, Donna J
   Principles of biochemical tests in diagnostic
microbiology.

   (Techniques in pure and applied microbiology) (A
Wiley biomedical publication)
   Includes bibliographical references and index.
   1. Bacteriology, Medical.   2. Microbiological
chemistry.   I. Ederer, Grace Mary, joint author.
II. Title.   III. Series.   [DNLM: 1. Biochemistry—
Laboratory manuals.   2. Bacteriology—Laboratory manuals.

QW25 B646p]
QR46.B53   616.07′581′01574192   75-17591
ISBN 0-471-08040-3

Printed in the United States of America

10 9 8 7 6 5 4 3 2

TO OUR STUDENTS

FOREWORD

It is frustrating to consult numerous sources without obtaining a full explanation of the chemical mechanisms at work in the diagnostic procedures used as everyday tools by the clinical microbiologist. The excellent manner in which Professors Blazevic and Ederer have collaborated to present not only an explanation of the principles in force, but also a compilation of the definitive literature cannot fail to bring a sigh of relief and an utterance of gratitude. This pertinent book will serve as a most valuable reference in my own professional library.

While working closely with the authors for eight years at the University of Minnesota I have been continually impressed with their endeavors to bring clarity and precision to the understanding of clinical microbiology. New and innovative teaching and technical methods have been implemented under their creative direction. Their efforts to outline, graphically illustrate, and record both routine and unusual features of microbial isolation and identification have considerably benefited laboratory personnel, trainees, and students. The content of this book is consistent with what this admirer has come to expect from these two highly respected colleagues.

The book is readable and easily understandable. It is sufficiently complete to be intellectually satisfying, accurate in its representations, valuable in its documentation and citation of the original literature, and delightful in the concise and orderly manner of presentation. The work is a very welcome addition to the printed armamentarium of the clinical microbiologist.

JOHN M. MATSEN, M.D.
Professor of Pathology and Pediatrics
Director of Clinical Laboratories
University of Utah Medical Center

May 1975

SERIES PREFACE

This series of handbooks is an effort to supply the practical advice that is needed in most laboratories active in the various fields of applied microbiology and to do it without an overdose of theoretical considerations. This however, does not imply that the books will be of only limited value to a theoretically oriented laboratory. Consider the extensive use of microorganisms as research tools now common among biophysicists, molecular biologists, immunologists, bioengineers and many others—and you will appreciate the need for a quick guide to the accepted techniques for handling bacteria and viruses.

Pure and applied microbiology go together; they are opposite sides of the same coin. The former is a road over forbiddingly steep hills on which the path is always partly hidden from view. The latter is the goal, for, after all, as Orville Wyss has emphasized, applied microbiology constitutes the backbone of our science, even if "we have responded to the gibes of the humanists who have always objected to the university leaving the cloister and entering the market place. It has never been demonstrated that the cloister is in any way superior to the market place for training a man to think, or that applied science is in any way inferior to pure science as an intellectual effort." There are many signs that the young student generation is more keenly aware of this than most of their professors, but this should not make the students forget Louis Pasteur's famous statement: "Without theory, practice is but routine born of habit. Theory alone can bring forth and develop the spirit of inventions." If the student keeps this in mind he will find that microbiology offers more challenging opportunities to make inventions that will affect man's future health and well-being than most other subjects which he might choose to study.

Carl-Göran Hedén

PREFACE

For out of olde feldes as men seyeth
Cometh all this newe corn fro yer to yere
And out of olde bokes, in good feyth
Cometh al this newe science that men lire

GEOFFREY CHAUCER, 1340–1400

In 1970 one of our clinical pathology residents asked us where he could find a book containing the basic principles behind the biochemical tests used in diagnostic bacteriology. Unfortunately, after much searching, we found that such a book did not exist nor could all of this information even be found in a wide assortment of texts on various aspects of bacteriology. Spurred on by the requests of other students we attempted to gather material on the biochemical tests useful in teaching. This led us, in many cases, back to the German literature of the late nineteenth and early twentieth century (renewing our admiration for such immortals as Voges, Proskauer, Ehrlich, and Kovacs). We learned that there really is not a great deal new in diagnostic bacteriology, but that much of what is known is buried in obscure and hard to find journals.

In the present book we attempt to bring together in one concise volume the biochemistry of the most common tests used in general diagnostic bacteriology from the aspect of both the bacteria and the test reactions. We do not include *all* tests ever used for identification of bacteria—only those that are generally well-accepted and suggested in current texts and manuals of diagnostic microbiology.

We express our special gratitude to Drs. Henry D. Isenberg, Dolph Klein, John C. Sherris, and Lee Anne McGonagle for their valuable suggestions and review of the manuscript.

DONNA J. BLAZEVIC
GRACE MARY EDERER

Minneapolis, Minnesota
May 1975

CONTENTS

PRINCIPLES OF BIOCHEMICAL TESTS IN DIAGNOSTIC MICROBIOLOGY

CHAPTER 1

ACETATE UTILIZATION

Some bacteria have the ability to use organic acids or the salts of these acids as a sole source of carbon for metabolism, and others do not. These differences in metabolic attributes have been particularly helpful in the differentiation of gram-negative bacilli belonging to the family Enterobacteriaceae. Specifically within this family the assessment of carbon utilization is very helpful in the differentiation of *Escherichia coli* and species of *Shigella; Shigella* cannot utilize acetate whereas most *E. coli* can use the carbon from this source.

Early investigators (1–3) recognized that some bacteria could multiply by utilizing organic acids instead of sugars as a source of carbon. In 1923 Koser (4) devised a basal medium composed of inorganic salts in which ammonium phosphate was an integral part of the formula to supply the nitrogen necessary for growth. To the basal medium Koser added various carbon sources encompassing a wide range of organic acids, including acetate. He evaluated the effect of the carbon sources on the growth of the coli-aerogenes group of organisms. Since coliform organisms grew equally well on basal medium containing acetate, this report about the usefulness of acetate aroused little interest at that time.

It was not until 1962 that Trabulsi and Ewing (5), in an effort to explore more completely the potential of organic acids to simplify the identification of Enterobacteriaceae, demonstrated that most *E. coli* could utilize sodium acetate, a sole source of carbon, whereas *Shigella* species could not. The medium they employed was essentially that of Simmons' citrate agar (6) with the citrate replaced by acetate.

When bacteria utilize the carbon of organic acids for growth, the end products of this metabolic process are carbonates, bicarbonates, and hydrogen, or sodium hydroxide. The production of alkaline products can be detected by an indicator such as bromthymol blue which is present in acetate medium as used by Trabulsi and Ewing (5). However, the detection of pH change is significant only if the medium does not contain sugars from which the organism could produce acid products or protein peptones, from which in turn it could produce alkaline products, leading to a false positive result for organic acid utilization. Therefore, the basal medium devised by Koser (4), modified by Simmons (6), and adapted by Trabulsi and Ewing (5) for assessing the utilization of acetate must be used to perform the test. When organisms are able to utilize the carbon of acetate, growth is observable in 1–7 days. As the organisms utilize acetate in their growth and break down the sodium acetate, the pH of the medium shifts from 6.8 to greater than 7.6, at which point the bromthymol blue indicator in the medium changes from green to blue.

The reactions that take place during the utilization of acetate are not yet completely understood. It is possible that acetate is oxidized to carbon dioxide and water (7). If so, as carbon dioxide accumulates during the reaction, carbonates would form and react with the sodium ions in the medium to form alkaline sodium carbonate. On the other hand, in utilizing the organic portion of the salts, the sodium ion may remain in excess. It may attract the hydroxyl group from available water in the medium, forming sodium hydroxide. Further research is needed to elucidate the mechanism of the alkaline reaction.

REFERENCES

1. Hoppe-Seyler, F. Üeber Gahrungsprozesse. *Z. Physiol. Chem.* **2:**1–28, 1879.

2. Ayers, H. and Rupp, P. The alkali-forming bacteria found in milk. *Science* **42:**318–319, 1915.

3. Pakes, W. C. C., and Jollyman, W. H. The bacterial decomposition of formic acid into carbon dioxide and hydrogen. *J. Chem. Soc. (London)* **79:**386–391, 1901.

4. Koser, S. A. Utilization of the salts of organic acids by the colon-aerogenes group. *J. Bact.* **8:**493–520, 1923.

5. Trabulsi, L. R. and Ewing, W. H., Sodium acetate medium for the differentiation of *Shigella* and *Escherichia* cultures. *Publ. Health Lab.* **20:**137–140, 1962.

6. Simmons, J. S. A culture medium for differentiating the organisms of the typhoid-colon aerogenes groups and for the isolation of certain fungi. *J. Infect Dis.* **39:**201–214, 1926.

7. Davis, B. D., Dulbecco, R., Eisen, H. N., Ginsberg, H. S., and Wood, W. B. *Microbiology* 2nd ed., Harper and Row, Hagerstown, Md. 1973, p. 50.

CHAPTER 2

ARGININE DIHYDROLASE

The detection of arginine dihydrolase activity of bacteria is helpful in the speciation of pseudomonads and *Enterobacter* as well as in the more complete characterization of beta hemolytic streptococci.

Horn (1) in 1933 was one of the first investigators to study breakdown products of arginine by bacteria and gave the name arginine desimidase to the enzyme that hydrolyzed arginine to citrulline. In 1940 Hills (2) reported that *Streptococcus hemolyticus* and *S. faecalis* possessed an enzyme for which he suggested the name arginine dihydrolase to distinguish it from arginase which acts upon arginine to yield ornithine and urea. According to Hills, arginine dihydrolase acted on arginine to yield ornithine, ammonia, and carbon dioxide without forming citrulline as an intermediate product. The reports of Horn and Hills were important in further elucidating the mechanism of action for arginine dihydrolase. In 1952 Slade and Slamp (3) showed that two enzymes were involved in the arginine dihydrolase system and suggested the names arginine desimidase and citrulline ureidase (citrullinase) for them. Arginine desimidase, which they showed could produce citrulline as an intermediate product, had already been named by Horn. Citrulline ureidase attacked the citrulline intermediate product producing ornithine, ammonia, and carbon dioxide. The stepwise reaction occurring in the decomposition of arginine was explored in more detail by other investigators (4–7). The reactions involved in the arginine dihydrolase system are shown below:

$$
\underset{\text{Arginine}}{\text{H}_2\text{N}-\overset{\overset{\displaystyle\text{NH}}{\|}}{\text{C}}-\text{N}-\overset{H}{\underset{H}{\text{C}}}-\overset{H}{\underset{H}{\text{C}}}-\overset{H}{\underset{H}{\text{C}}}-\overset{H}{\underset{\text{NH}_2}{\text{C}}}-\text{COOH}}
\quad\xrightarrow[\text{desimidase}]{\text{arginine}}
$$

$$
\underset{\text{Citrulline}}{\text{H}_2\text{N}-\overset{\overset{\displaystyle\text{O}}{\|}}{\text{C}}-\text{N}-\overset{H}{\underset{H}{\text{C}}}-\overset{H}{\underset{H}{\text{C}}}-\overset{H}{\underset{H}{\text{C}}}-\overset{H}{\underset{\text{NH}_2}{\text{C}}}-\text{COOH} + \text{NH}_3}
\quad\xrightarrow[\text{(citrullinase)}]{\text{citrulline ureidase}}
$$

$$H_2N-\overset{\displaystyle H}{\underset{\displaystyle H}{C}}-\overset{\displaystyle H}{\underset{\displaystyle H}{C}}-\overset{\displaystyle H}{\underset{\displaystyle H}{C}}-\overset{\displaystyle H}{\underset{\displaystyle NH_2}{C}}-COOH + NH_3 + CO_2$$

Ornithine

The early definitive work on arginine dihydrolase was carried out using gasometric and titrametric methods which were not suitable for clinical use. Since one of the end products of the action of arginine dihydrolase is ammonia, several media were developed for clinical usage. The enzyme was detected by a pH shift from acid to alkaline using an indicator. A method was developed by Moeller (8) in 1955 in which 1% L-arginine monohydrochloride was added to a basal medium composed of glucose, peptones, beef extract, and pyridoxal with a combined indicator system—bromcresol purple and cresol red. The organisms were inoculated into the media, covered with sterile mineral oil, and incubated at 37°C for 6 days. A tube of medium without arginine was used for a control. The medium worked well with members of the Enterobacteriaceae, since the glucose was first fermented and the indicator showed a pH shift to acid; when arginine was attacked by arginine dihydrolase the pH of the medium shifted from acid to alkaline. The arginine dihydrolase activity of nonfermenters was more difficult to assess using Moeller's medium, although Sherris and associates (9) stated that nonfermenters negative for arginine dihydrolase produced a slate grey color in the medium whereas those that were positive for this enzyme changed the indicator to a violet color within 24 hours.

While studying the usefulness of detecting arginine dihydrolase activity in the pseudomonads Sherris and his colleagues (9) found that they could obviate the need for a paraffin or mineral oil seal by using more substrate in a tube of smaller diameter, thus diminishing the surface exposed to air and subsequent oxidative alkalinization. Taylor and Whitby (10) evaluated a medium developed by Thornley (11) which contained L-arginine monohydrochloride, peptone, salts, a small amount of agar, and phenol red for an indicator. This medium did not contain glucose or pyridoxal. They found this medium to be very useful for further characterization of fermenters and nonfermenters; the ammonia produced caused the indicator to change to alkaline (deep pink).

Although media that detected pH changes were helpful from a clinical viewpoint, the need to know the exact products involved in the reaction still remained. With the advent of paper chromatography it was possible to detect amino acid products simply with very little equipment. Using paper chromatography, mixtures of amino acids can be separated because of differences in partition coefficients of various amino acids between an organic and an aqueous phase. Soru (12) in 1950 working with circular paper chromatography was able to detect arginine dihydrolase activity in hemolytic streptococci. Since arginine, ornithine, and citrulline differed in solubility in organic solvent, each moved at a different rate on the paper, thus showing distinctly separated spots when sprayed with ninhydrin.

Although the use of paper chromatography provided definitive results concerning the metabolism of arginine, the procedure was lengthy, requiring several days. Thin-layer chromatography (TLC) is similar in principle to paper chromatography; it provides much more rapid results and achieves reproducible separation of amino acids as well as other organic substances. In our laboratory we developed a TLC method for determining arginine dihydrolase activity in nonfermentative organisms (13). Using cellulose precoated sheets we were able to detect arginine dihydrolase activity in these organisms in a few hours. We have also found this rapid TLC method useful for speciating *Enterobacter* (14). Other investigators (15) have used TLC in detecting arginine dihydrolase activity and have recommended modification of the solvent system to improve the method's sensitivity.

When performing arginine dihydrolase tests using an arginine basal medium with an indicator it is important to (*a*) inoculate a control tube without arginine and (*b*) seal each tube with paraffin or sterile mineral oil. The pH of the control tube should remain unchanged or shift to acid, depending on the method, and remain at that pH. Should the control tube become alkaline, the test must be repeated. In narrow tubes the mineral oil or paraffin seal may be omitted. However, at the top of the medium some alkalinity will be observed because of oxidative metabolism of peptones. A seal is not necessary when using definitive methods such as TLC or other chromatographic procedures, since the arginine substrate and amino acid products are specifically identified.

REFERENCES

1. Horn, F. Über den Abbau des Arginins zu Citrullin durch *Bacillus pyocyaneous*. *Z. Physiol. Chem.* **216**:244–247, 1933.

2. Hills, G. M. Ammonia production by pathogenic bacteria. *Biochem. J.* **34**:1057–1069, 1940.

3. Slade, H. D. and Slamp, W. C. The formation of arginine dihydrolase by streptococci and some properties of the enzyme system. *J. Bact.* **64**:455–466, 1952.

4. Oginsky, E. L. and Gehrig, R. F. The arginine dihydrolase system of *Streptococcus faecalis*: Identification of citrulline as an intermediate product. *J. Biol. Chem.* **198**:791–797, 1952.

5. Knivett, V. A. Citrulline as an intermediate in the breakdown of arginine by *Streptococcus faecalis*. *Biochem. J.* **50**:xxx–xxxi, 1952.

6. Oginsky, E. L. and Gehrig. R. F. The arginine dihydrolase system of *Streptococcus faecalis:* The decomposition of citrulline. *J. Biol. Chem.* **204**: 721–729, 1953.

7. Oginsky, E. L. Arginine dihydrolase. In S. P. Colowick and N. O. Kaplan. *Methods in Enzymology*, Vol. 2. Academic Press, New York, 1955.

8. Moeller, V. Simplified tests for some amino acid decarboxylases and for the arginine dihydrolase system. *Acta Pathol. Microbiol. Scand.* **36**:158–172, 1955.

9. Sherris, J. C., Shoesmith, J. G., Parker, M. T., and Breckon, D. Tests for the rapid breakdown of arginine by bacteria: Their use in the identification of pseudomonads. *J. Gen. Microbiol.* **21**:389–396, 1959.

10. Taylor, J. J. and Whitby, J. L. *Pseudomonas pyocyanea* and the arginine dihydrolase system. *J. Clin. Pathol.* **17**:122–125, 1964.

11. Thornley, M. J. The differentiation of *Pseudomonas* from other gram-negative bacteria on the basis of arginine metabolism. *J. Appl. Bact.* **23**:37–52, 1960.

12. Soru, E. Application of circular paper chromatography to the differentiation of bacteria by enzymic tests. *J. Chromatogr.* **1**:380–384, 1958.

13. Williams, G. A., Blazevic, D. J., and Ederer, G. M. Detection of arginine dihydrolase in nonfermentative bacteria by use of thin-layer chromatography. *Appl. Microbiol.* **22**:1135–1137, 1971.

14. Blazevic, D. J. and Ederer, G. M. Dectection of arginine dihydrolase activity in *Enterobacter* by the use of thin-layer chromatography. Abstracts II, First International Congress for Bacteriology. 1973, p. 133.

15. Zolg, W. and Ottow, J. C. G. Improved thin-layer technique for detection of arginine dihydrolase among the *Pseudomonas* species. *Appl. Microbiol.* **26**:1001–1003, 1973.

CHAPTER 3

BILE SOLUBILITY

The bile solubility test is used chiefly to differentiate pneumococci from alpha hemolytic streptococci.

In 1900 Neufeld (1) observed that pneumococci were soluble in bile whereas streptococci were not. Since the pneumococcus is also subject to autolysis in culture, the exact nature of bile solubility and autolysis has been investigated by many scientists. Avery and Cullen (2) in 1923 showed that pneumococci possess an intracellular lytic enzyme which produced lysis of heat-killed pneumococci of the same type or heterologous types. These same heat-killed cultures of pneumococci were impervious to lysis by bile, which indicated that the autolytic heat labile enzyme must be present for the solution of the organisms by bile. A few years later Mair (3) reported that bile merely accelerated the natural autolytic process. Twenty bile salts were evaluated by Downie, Stent, and White (4) for their ability to lyse pneumococci. It was found the sodium salt of deoxycholic acid produced the most rapid lysis, even in very low concentrations. The authors thought there was a correlation between the chemical structure of the bile acid from which the salt was derived, its ability to form additional compounds, and the lytic activity of the substance. They felt that the lytic activity was related to the number and position of the hydroxyl groups. Bile salts that did not have a hydroxyl group, such as dehydrocholic acid and dehydrodeoxycholic acid, were inactive. The common carbon structure and the position of the hydroxyl group in one of the compounds, studied by Downie and associates (4) are shown below.

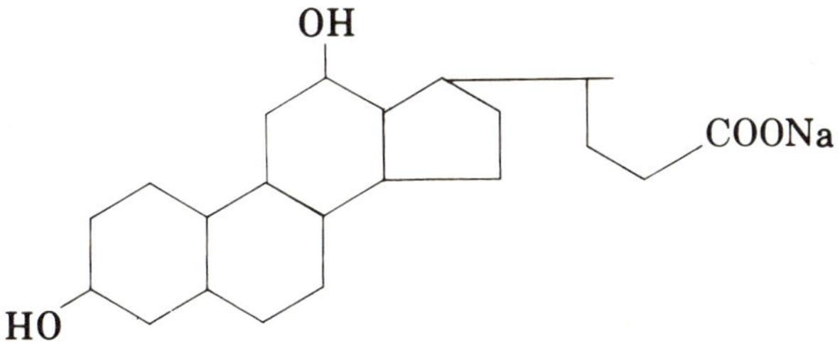

Sodium deoxycholate

The solubility of pneumococci in saponin had been previously described (4) but was more completely investigated by Klein (5, 6). He had to sensitize the pneumococci by growing the organisms in media containing blood, serum, ascitic fluid, or cholesterol before the pneumococci could be lysed by saponin. He found the saponin solubility test correlated well with bile solubility tests and was less expensive. The saponin solubility test, however, did not gain wide usage. It is interesting to note that the chemical structures of saponins are similar to that of bile salts, as shown below; saponins also have a hydroxyl group at the carbon 3 position.

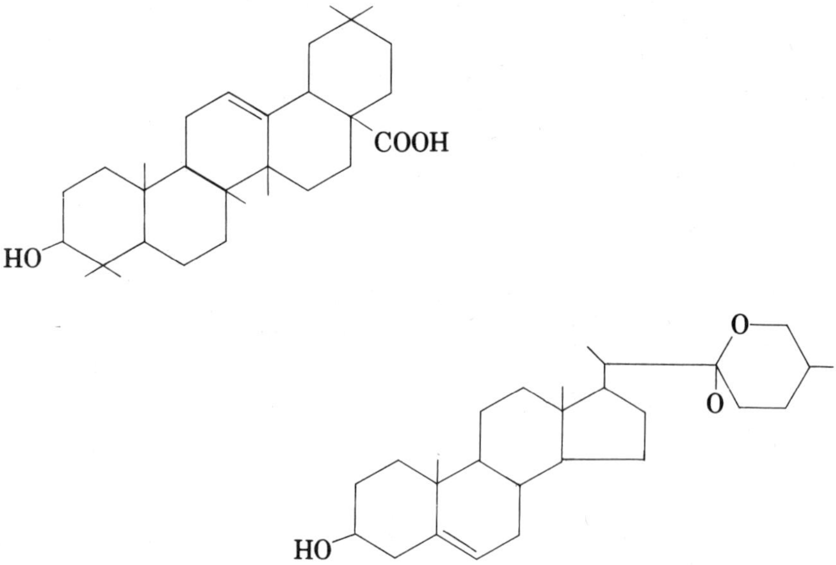

Saponins

In 1965 Hawn and Beebe (7) showed that bile solubility testing could be performed with accuracy directly on the primary blood agar, plate, therefore simplifying the procedure. They used a 2% deoxycholate solution, added a loopful of this agent to the suspect colony, incubated it at 37°C for 30 minutes, and examined the site with a hand lens to determine whether the colony had dissolved. Although these authors remark that sodium lauryl sulfate is often used for this test, it is "apparently" less specific than sodium deoxycholate. In our laboratories we have applied the principle of colonial bile solubility testing (8) but have successfully used a 1% solution of sodium lauryl sulfate under the trade name of Dreft. Currently recommended agents and concentrations for bile solubility tube tests are 2% bile salts (9) and 10% sodium deoxycholate (10).

During the years when the best agent for bile solubility testing of pneumococci was being investigated other pertinent aspects of the test were also considered. As early as 1917 Mair (11) recognized that it was necessary to alkalinize acidic cultures of pneumococci to prevent the bile acids from precipitating out and obscuring reading the test. Avery and Cullen (2) found that the optimum pH range for lysis was 6–8. Serum was often added to broth to regulate the pH, and Holt (12) suggested that sodium carbonate was equally effective and less expensive. He devised a basal medium containing cysteine. The cysteine provided protection against peroxide formation of the organism by making the medium anaerobic. To the basal medium he added sterile glucose and sodium carbonate and achieved luxurious growth of pneumococci. This medium, though effective, was too cumbersome to prepare for routine use. Falk and Yang (13) showed that chlorides with monovalent cations at relatively low concentrations inhibit the solution of pneumococci by bile but at higher concentrations may accelerate the lytic action. They found that chlorides with divalent cations behaved in the opposite way.

During the pre-antimicrobial agent period scientists and physicians explored the efficacy of *in vivo* injection of bile salts into the previously aspirated abscess cavity of patients with pneumococcal empyema. Anderson and Hart (14), in relationship to minimizing the potential toxic effects of such treatment, evaluated the lowest concentration of sodium chloride and deoxycholate which would produce lysis of pneumococci in 1 hour. They found that very low concentrations of deoxycholate would produce lysis of pneumococci when the concentration of sodium chloride was reduced to 0.4%. Lysis of pneumococci could also be achieved in the absence of sodium chloride but higher concentrations of deoxycholate were required to produce lysis. By varying the concentrations of sodium chloride and deoxycholate they demonstrated the reciprocal relation between these two substances and lysis of pneumococci.

Basic research concerning the mechanism of action involved in the lysis of pneumococci diminished with the advent of sulfonamide and antibiotic therapy. Recently, however, Masser and Tomasz (15), using crude extracts of pneumococcal cellular contents, were able to show that the autolytic enzyme present in these organisms appears to be an amidase which splits the muramic acid and alanine bond in the peptidoglycan portion of the cell wall. They found the action of this amidase to be dependent on the choline component of the techoic acid in the cell wall; if ethanolamine was substituted for choline in the cell wall, the cells were resistant to lysis by this enzyme. Since these investigators were working with a crude enzyme extract they thought there was a possibility that more than one enzyme might be responsible for autolysis.

When performing solubility testing on pneumococci in the diagnostic microbiology laboratory today, we should keep in mind the reports of early investigators concerning certain variables that might affect the validity of the results. It is well to remember, for example, to use young viable cultures of pneumococci for solubility testing, since dead organisms do not lyse. Equivocal results may be encountered when older cultures are employed. On the other hand, pneumococci may autolyse, thus causing the loss of a culture. The pH of the broth culture of pneumococci to be tested should be in a range of 6.8–7.6. Precipitates occur if bile salts or deoxycholate are added to an acid broth culture. The reagents used in the test should be in the proper proportions to achieve accurate results. The regular evaluation of all materials in determining the solubility of pneumococci should be a part of the quality control program of each laboratory. Finally it should be kept in mind that other organisms, such as *Haemophilus*, can be lysed by deoxycholate or bile salts (16). Therefore, the genus of the suspect organism must be determined accurately before solubility testing is performed.

REFERENCES

1. Neufeld, F. Über eine specifishe bakteriolytische Wirkung der Galle. *Z. Hyg. Infektionskr.* **34:**454–464, 1900.

2. Avery, O. T. and Cullen, G. E. Studies on the enzymes of pneumococcus. *J. Exp. Med.* **38:**199–206, 1923.

3. Mair, W. A system of bacteriology. *Med. Res. Counc., London,* **2:**168, 1929.

4. Downie, A. W., Stent, L., and White, S. M. The bile solubility of pneumococcus with special reference to the chemical structure of various bile-salts. *Br. J. Exp. Pathol.* **12:**1–9, 1931.

5. Klein, S. J. Studies on the solubility of pneumococcus in saponin; sensitization by ergosterol. *J. Bact.* **26:**215–219, 1933.

6. Klein, S. J. Studies on the solubility of pneumococcus in saponin; the saponin-lysis reaction as a means of differentiating pneumococcus and streptococcus. *J. Bact.* **30:**43–48, 1935.

7. Hawn, V. Z. and Beebe, E. Rapid method for demonstrating bile solubility of *Diplococcus pneumoniae. J. Bact.* **60:**549, 1965.

8. Blazevic, D. J. *Laboratory Procedures in Diagnostic Microbiology.* Telstar Productions, Inc., St. Paul, Minn. 1972.

9. Blair, E. B. Media, test procedures, and chemical reagents. In H. L. Bodily, E. L. Updyke, and J. O. Mason (Eds.). *Diagnostic Procedures for Bacterial, Mycotic and Parastic Infections,* 5th ed. American Public Health Association, New York, 1970.

10. Austrian, R. Diplococcus pneumoniae In J. E. Blair, E. H. Lennette, and J. P. Truant (Eds.). *Manual of Clinical Microbiology*. American Society of Microbiology, Bethesda, Md. 1970.

11. Mair, W. A contribution to the serological classification of the bile-soluble diplococci. *J. Pathol Bact.* **21**:305–328, 1917.

12. Holt, L. B. The culture of *Streptococcus pneumoniae, J. Gen. Microbiol.* **27**:327–330, 1962.

13. Falk, I. S. and Yang, S. Y. Studies of respiratory diseases: The influence of certain electrolytes and non-electrolytes on the bile solubility of pneumococci. *J. Infect. Dis.* **38**:1–7, 1926.

14. Anderson, A. B. and Hart, P. D. The lysis of pneumococci by sodium deoxycholate. *Lancet* 359–360, 1934.

15. Mosser, J. L., and Tomasz, A. Choline-containing techoic acid as a structural component of pneumococcal cell wall and its role in sensitivity to lysis by an autolytic enzyme. *J. Biol. Chem.* **245**:287–298, 1970.

16. Eveland, W. C. *Haemophilus* infections. In H. L. Bodily, E. L. Updyke, and J. O. Mason (Eds.). *Diagnostic Procedures for Bacterial, Mycotic and Parasitic Infections,* 5th ed. American Public Health Association, New York 1970, p. 131.

CHAPTER 4

CATALASE

The catalase test has been routinely used for many years as an aid in distinguishing between staphylococci and streptococci and in the identification of other bacteria.

McLeod and Gordon (1) in 1923 were the first investigators to propose that the detection of this enzyme might be useful in identifying bacteria. They divided bacteria into classes according to their ability to produce hydrogen peroxide and catalase. The close connection between the inability of organisms to produce catalase and the fact that the same organisms tended to die off on subculture was noted. They suggested that these catalase-nonproducing bacteria tended to die in sugar-free media either because they produced hydrogen peroxide or because hydrogen peroxide accumulated in media when it was exposed to air and light. Their recommendations that stock cultures be stored where the access to air is limited and in the dark are still practical.

Zeile and Hellstrom (2) first isolated the enzyme catalase from horse liver in 1930. These investigators characterized catalase as a haematin compound. They also isolated catalase from germinating cucumber seeds.

In 1938 Keilin and Hartree (3) confirmed this earlier work (2) and also found that catalase could be prepared from red blood cells. Using a horse liver catalase preparation Keilin and Hartree explained the mechanism of action of catalase and hydrogen peroxide as follows:

1. Anaerobic reduction of catalase iron; $4 \ Fe^{3+} + 2 \ H_2O_2 \rightarrow 4 \ Fe^{2+} + 4 \ H^+ + 2O_2$.
2. Aerobic oxidation by O_2; $4 \ Fe^{2+} + 4 \ H^+ + O_2 \rightarrow 4 \ Fe^{3+} + 2H_2O$.

With these principles in mind, it becomes clear why catalase tests on anaerobic bacteria should be performed only after exposing the organisms to the air for approximately ½ hour.

The chemical reaction that occurs in the performance of the catalase test is expressed simply in the following equation:

$$2 \text{ H}_2\text{O}_2 \xrightarrow{\text{catalase}} 2 \text{ H}_2\text{O} + \text{O}_2 \uparrow$$

The production of oxygen gas bubbles, in other words, has been the criterion
for a positive test. Until 1972 little attention was directed toward the standardization of the catalase test. At this time Taylor and Achanzar (4) reviewed
the references for catalase test and found, for example, a wide range of hydrogen peroxide concentration used in the test—1.5–30%. They evaluated the
efficacy of hydrogen peroxide strengths from 0.5 to 5.0% and concluded that
one drop of a 3% concentration was the most effective. They found that the test
could be performed equally well on an isolated colony grown on blood-non-
containing medium (3), in broth, or by taking the colony to a microscope slide
with a coverslip. They emphasized that catalase tests must not be performed on
cultures older than 24 hours, since the enzyme is present only in viable cultures
and false negatives can occur in old cultures. Using these controlled measures,
they divided the Enterobacteriaceae into genera, which have reactions with hydrogen peroxide through describable degrees—vigorous, moderate, weak, and
negative. Standardization and quality control of the procedure are prime requirements for the interpretation of this simple test.

REFERENCES

1. McLeod, M. B. and Gordon, J. Catalase production and sensitiveness to hydrogen peroxide
    amongst bacteria: With a scheme of classification based on these principles. *J. Pathol. Bact.*
    26: 326–331, 1923.

2. Zeile, K. and Hellstrom, H. Über die aktive Gruppe der Leberkatalase. *Hoppe-Seyler Z.*
    192: 171–192, 1930.

3. Keilin, D. and Hartree, E. F. On the mechanism of decompositon of hydrogen peroxide by
    catalase. *Proc. Roy. Soc. B.* **124:** 397–405, 1938.

4. Taylor, W. I. and Achanzar, D. Catalase test as an aid in the identification of Enterobac-
    teriaceae. *Appl. Microbiol.* **24:** 58–61, 1972.

CHAPTER 5

CITRATE UTILIZATION

The ability of certain microorganisms to utilize the carbon of sodium citrate as the sole source of carbon has been used chiefly in the differentiation of gram-negative bacilli.

Hoppe-Seyler (1) in 1879 was the first investigator to suggest that the salts of organic acids could be converted to alkaline carbonates by bacteria. The application and further elucidation of this significant finding was not pursued until 1915 when Ayers and Rupp (2) observed that certain bacteria could be defined as "alkali-producing bacteria" if they produced an alkaline reaction in milk without peptonizing it. The authors thought that the alkaline reaction was caused by the breakdown of salts of organic acids in the milk to carbonates and "that fermentation of the salts of organic acids will be of great value in the classification of bacteria. . . ."

During the next few years Ayers and Rupp (3) focused their research on alkaline reactions produced by certain bacteria which could not be explained by the production of ammonia or basic substances from protein decomposition. To make sure that alkaline products were not formed by bacteria breaking down peptone, they supplied the nitrogen needed for growth by using a sodium ammonium phosphate medium—a peptone-free nitrogen base. They evaluated the effect of two carbon sources used separately and in a mixture, namely dextrose and sodium citrate. They concluded that acid and alkaline fermentations could be produced simultaneously through fermentation of sugar and salts or organic acids and that the rate of fermentation of dextrose and citrate separately varied with different bacteria.

Brown (4) explored the use of sodium citrate in media still further and found that the growth of certain organisms was enhanced by this organic acid, whereas the growth of others was completely inhibited. He was also able to show that citrate added to liquid or solid medium produced the same bacterial growth patterns. In 1923 Koser (5) devised a liquid basal broth medium in which ammonium phosphate supplied the nitrogen, to which he added each organic acid in a concentration of 0.2%. He studied 18 organic acids and evaluated the ability of various organisms to utilize these acids as carbon sources. The most striking result of his work was the recognition that the aerogenes group of organisms can utilize citrate for growth and that coli organisms are completely unable to grow in this medium. This was the introduction of carbon utilization as a diagnostic aid. Since determining growth on solid medium is easier than interpreting turbidity in broth, Simmons (6) added agar to Koser's medium and incorporated an indicator, bromthymol blue, which was green at a pH of 6.8 and blue at a pH greater than 7.6.

Another medium to demonstrate citrate utilization was developed by Christensen (7) in which the nitrogen was supplied by cysteine hydrochloride. Yeast extract was added for enrichment and phenol red for the indicator. This medium is rich in nutrients, and organisms may be positive on this medium and negative on Simmons' medium. A test for citrate utilization has also been developed using citrate impregnated strips. The correlation of this test method with Simmons' citrate is poor (8).

In 1947 Lominski and associates (9) showed that although *Escherichia coli* failed to grow in Koser's citrate medium the organism was capable of utilizing citrate under suitable conditions, especially when glucose was available in the medium as well as citrate. These investigators felt that growth of *E. coli* on citrate medium depended on having an appropriate hydrogen acceptor in the medium.

Dagley and Dawes (10) showed that citrate could be dissimilated by an enzyme extracted from *Aerobacter aerogenes* (*Enterobacter*) into oxaloacetate and acetate followed by decarboxylation of oxaloacetate to pyruvate and carbon dioxide. They found that the presence of magnesium ion was necessary for maximum reactivity of the enzyme. From the findings of these investigators it would seem that an organism either does or does not posses an enzyme that can catalyze the utilization of citrate. This theory was disproved by Wheat and Ajl (11) who showed that disrupted cells of *E. coli* displayed enzymatic activity specific for the breakdown of citrate. They were also able to show that the citrate splitting enzyme of *E. coli* was adaptive and was formed in large amounts in the presence of both glucose and citrate.

Recent studies of O'Brien and Geisler (12) indicate that there are differences in citrate metabolism even between species of the same genus. They found, for example, that enzyme activity by *A. cloacae* (*Enterobacter*) did not require magnesium nor was it stimulated by this ion.

The enzyme that catalyzes the cleavage of citrate has been given a number of names: citrate lyase, citritase, citrase, citrate aldolase, citridesmolase. The primary products of cleavage are oxaloacetate and acetate which are subsequently converted to pyruvate and carbon dioxide by an oxaloacetate decarboxylase. The reaction may or may not require a divalent metal ion such as manganese or magnesium (10, 12, 13). Since *E. coli* has been shown to possess citrate lyase although it cannot utilize the citrate in Koser's or Simmons' media, it apparently lacks the transport system or permease that would permit the citrate to enter the cell to be used as a carbon source (14). The permease for citrate utilization is inducible.

The exact nature of the alkaline reaction produced by organisms that utilize the carbon of citrate for growth is not totally clear even at this time. Although an ammonium salt is present for the nitrogen source Clark and Lubs (15) as early as 1917 very clearly demonstrated that the alkaline reaction involved in citrate utilization was not due to the release of ammonia. It appears, therefore, that the alkaline reaction most likely occurs because an excess of carbon dioxide is generated as citrate is cleaved to oxaloacetate, which is then decarboxylated to pyruvate and carbon dioxide. The excess of carbon dioxide may combine with sodium and water to form sodium carbonate, which is sufficiently alkaline to change the indicator, bromthymol blue, from green to deep blue. It is also possible that the sodium ion may remain in excess in this reaction and may attract the hydroxyl group from the available water in the medium to form sodium hydroxide.

When performing tests to determine citrate utilization by bacteria, it is important to use a light inoculum and minimize the carry-over of any nutritive constituents of the culture medium from which the inoculum is taken. Incubation times vary, depending on the method for reporting negative results, from 4 hours for the impregnated paper strip to 4–7 days for Simmons' citrate. Questionable positive cultural reactions should be confirmed by subculture to fresh citrate medium. When using citrate medium with an indicator system incorporated into it, only those cultures that demonstrate a pH shift to alkaline should be considered positive.

REFERENCES

1. Hoppe-Seyler, F. Über Gahrungsprozesse. *Z. Physiol. Chem.* **2**:1–28, 1879.
2. Ayers, S. H. and Rupp, P. Alkali-forming bacteria in milk. *Science* **42**:318–319, 1915.

3. Ayers, S. H. and Rupp, P. Simultaneous acid and alkaline bacterial fermentations from dextrose and salts of organic acids respectively. *J. Infect. Dis.* **23**:188–216, 1918.

4. Brown, H. C. Observations on the use of citrated media. *Lancet* No. 5079, 22–23, 1921.

5. Koser, S. A. Utilization of the salts of organic acids by the colon-aerogenes group *J. Bact.* **8**:493–520, 1923.

6. Simmons, J. S. A culture medium for differentiating organisms of the typhoid-colon aerogenes groups and for isolation of certain fungi. *J. Infect. Dis.* **39**:209–214, 1926.

7. Christensen, W. B. Hydrogen sulfide production and citrate utilization in the differentiation of enteric pathogens and coliform bacteria. *Res. Bull. Weld County Health Dept., Greeley, Colo.* **1**:3, 1949.

8. Matsen, J. M. and Sherris, J. C. Comparative study of the efficacy of seven paper-impregnated strips and conventional biochemical tests in identifying gram-negative organisms. *Appl. Microbiol.* **18**:452–457, 1969.

9. Lominski, I., Conway, N. S., Harper, E. M., and Rennie, J. B. Utilization of citric acid by some so-called citrate-non-utilizing bacteria. *Nature* **160**:573–574, 1947.

10. Dagley, S. and Dawes, E. A. Dissimilation of citric acid by bacterial extracts. *Nature* **172**:345–346, 1956.

11. Wheat, R. W. and Ajl, S. J. Citritase the citrate-splitting enzyme from *Escherichia coli. J. Biol. Chem.* **217**:897–920, 1955.

12. O'Brien, R. W. and Geisler, J. Citrate metabolism in *Aerobacter cloacae. J. Bact.* **119**:661–665, 1974.

13. Reeves, H. C. Rabin, R., Wegener, W. S. and Ajl, S. J. In *Methods in Microbiology*, Vol. 6A. Norris, J. R. and Ribbons, D. W., Eds. Academic Press, London and New York, 1971, Chapter 10, pp. 430–432.

14. Mandelstam, J. and McQuillen, K. *Biochemistry of Bacterial Growth*. John Wiley and Sons, New York, 1968, pp. 424–426.

15. Clark, W. M. and Lubs, H. A. Improved chemical methods for differentiating bacteria of the coli-aerogenes family. *J. Biol. Chem.* **30**:209–234, 1917.

CHAPTER 6

COAGULASE

The test for coagulase production by staphylococci has become the most accurate single criterion used in diagnostic microbiology laboratories to differentiate pathogenic from nonpathogenic staphylococci. It is the former that produce coagulase.

Loeb (1) in 1903 was the first investigator to observe that some bacteria were able to clot blood plasma; pyogenic staphylococci demonstrated stronger coagulase activity than any other bacteria tested. Cadness-Graves and associates (2) in 1943 did the first definitive study evaluating the correlation between coagulase tests performed by the tube method and those using the slide procedure; they found the correlation to be good. They introduced the term "clumping" for the resultant positive slide reaction, and recommended that a tube coagulase test be performed on all negative slide tests.

The first investigators to shed light on the mechanism of action of coagulase or staphylocoagulase, as it is often called, were Smith and Hale (3) in 1944. They clarified several important points in the understanding of the basic aspects of the reaction. (*a*) Coagulase is thermostable and filterable. (*b*) Coagulase is a precursor of a thrombin-like substance that depends on an activator present in the plasma of many species but absent in others (mouse, guinea pig, and fowl). (*c*) The thrombin-like substance is thermolabile and its concentration at any moment is most dependent on temperature, time, and concentration of reagents. (*d*) Plasmas of those species that are not coagulated by coagulase are deficient in activator. (*e*) The clotting ability of coagulase is not dependent on calcium ions.

The mechanism of action of coagulase was further elucidated in 1954 by Duthrie (4). He identified two forms of coagulase: one bound to the cell wall and one liberated by the cell as free coagulase. He attributed the slide clumping reaction to bound coagulase which acted directly on the fibrinogen of certain animals whereas the free coagulase acted on the prothrombin to give a thrombin-like product. He found these two coagulases to be distinctly different antigenically.

The exact mechanism of coagulase is not yet completely understood. It has been generally agreed that a plasma factor is involved in the reaction and that this activator factor resembles prothrombin. A simplified reaction scheme is shown below (5, 6).

Coagulase + coagulase reacting factor (CRF) → coagulase-thrombin (CT)
Coagulase-thrombin (CT) + fibrinogen → fibrin (clot)

Although Murray and Ghodes (7) proposed that CRF plays only a secondary role in the coagulase reaction acting as an accelerator, all investigators in this field seem to agree that the final target for coagulase action is fibrinogen.

According to the kinetic studies of Soulier and associates (8) staphylocoagulase appears to act upon prothrombin by forming a molecular complex with it rather than by altering prothrombin enzymatically. Work such as this may well reveal that the mechanism of action of staphylocoagulase could be described quite differently than is shown in the preceding paragraph.

In 1974 Tager (6) summarized the current views on mechanisms of coagulase action and the relationship of coagulase clotting of blood to physiological clotting involving prothrombin and other factors necessary for physiological clotting of blood. He reviewed the differences that have been shown for coagulase clotting as opposed to physiological clotting. A few of the major differences are listed below:

1.   The clotting reaction with coagulase does not require Factors V, VII, VIII, X, Xa, XI, and XII. However, plasmas deficient in Factor II, prothrombin, have a reduced capacity to clot with coagulase.

2.   Coagulase clotting does not require calcium ions, since the reaction occurs when citrate, oxalate, fluoride, and ethylenediaminetetraacetate (EDTA) are present. Neither hirudin nor heparin interfere with coagulase clotting.

When performing tube coagulase tests on staphylococci in the clinical laboratory in which citrated plasma is used, it is important to keep in mind that any organism that can utilize citrate will produce a clot (9). It is of paramount importance, therefore, to identify the organism as a staphylococcus and to be sure that it is in pure culture before performing a tube coagulase test. One must be aware also that some human blood, should outdated human bank blood be used for the test, contains inhibitory factors (10). Good quality control measures for this test, using a combination of strong and weak positive as well as a negative control organism, will provide assurance for the validity of the test.

REFERENCES

1. Loeb, L. The influence of certain bacteria on the coagulation of the blood. *J. Med. Res.* **10**:407–419, 1903.

2. Cadness-Graves, B., Williams, R., Harper, G. J., and Miles, A. A. Slide-test for coagulase-positive staphylococci. *Lancet* **i,** 736–738, 1943.

3. Smith, W. and Hale, J. H. The nature and mode of action of staphylococcus coagulase. *Br. J. Exp. Pathol.* **25**:101–110, 1944.

4. Duthrie, E. S. Evidence of two forms of staphylococcal coagulase. *J. Gen. Microbiol.* **10**:427–436, 1954.

5. Tager, M. and Drummon, M. C. Implications of staphylocoagulase-thrombin and fibrinogen interaction. *Thromb. Diath. Haematol.* **23**:291–298, 1970.

6. Tager, M. Current views on the mechanisms of coagulase action in blood clotting. *In* Recent Advances in Staphylococcal Research. *Ann. N. Y. Acad. Sci.* **236**:277–291, 1974.

7. Murray, M. P., and Ghodes, P. Role of coagulase reacting factor in the activation of fibrinogen. *J. Bact.* **78**:450–451, 1959.

8. Soulier, J. P., Prou, O., Halle, L. Further studies on thrombin-coagulase. *Thromb. Diath. Haematol.* **23**:37–39, 1970.

9. Paik, G. and Suggs, M. T. Reagents, stains and miscellaneous test procedures. In Lennette, E. H., Spaulding, E. H., and Truant, J. P., Eds. *Manual of Clinical Microbiology,* 2nd ed. American Society for Microbiolgy, Bethesda, Md., 1974, p. 931.

10. Bailey, W. R. and Scott, E. G. *Diagnostic Microbiology,* 4th ed. The C. V. Mosby Co., St. Louis, Mo., 1974, p. 112.

CHAPTER 7

DEAMINASE TESTS

The ability of *Proteus* and *Providence* species to deaminate a number of amino acids provides a means of distinguishing these genera from other members of the family Enterobacteriaceae.

In the early part of the twentieth century many investigators were studying the metabolism af amino acids by bacteria. Choosing a strain of *Proteus* Bernheim and associates (1) in 1935 studied the ability of this organism to oxidize amino acids and found that all natural optical isomers of more than 10 amino acids studied were oxidized by this organism. They stated that oxidation of an amino acid corresponded to deamination of that amino acid. For their experiments they used a phosphate buffered suspension of an 18-hour culture of the organism from a beef extract agar slant which they described as a "resting" culture of *Proteus*. The suspensions of organisms were added to each of the amino acids dissolved in phosphate buffer, and the rate and amount of oxygen uptake was measured using a Warburg vessel. The optimum pH for oxidation of the amino acids was found to be between 7.2 and 8.3.

Stumpf and Green (2) in 1944 also used gasometric methods to study the oxidative enzyme of *Proteus vulgaris* more completely with the aid of intact bacterial cells as well as cell-free extracts. They named the enzyme L-amino acid oxidase, a name that is still in use. These investigators found that 11 of the 22 amino acids tested were deaminated by amino acid oxidase and that the most rapid reaction was with phenylalanine. They noted that the velocity of the deaminase reaction was related to the number of carbon atoms in the unbranched amino acids; amino acids with six carbon atoms demonstrated the greatest velocity. In the phenyl-substituted series the velocity of the reaction was found to be related to the number of carbon atoms in the aliphatic side chain; phenylalanine with three carbons in the side chain was oxidized most rapidly. Amino acids with fewer carbons were not oxidized by the enzyme and those with more carbons were oxidized at a markedly slower rate. The general equation for the oxidation of amino acids by amino acid oxidase proposed by these investigators (2) is

$$RCHNH_2COOH + \tfrac{1}{2}\,O_2 \rightarrow RCOCOOH + NH_3 + H_2O$$

To confirm the validity of this equation, Stumpf and Green (2) made hydrazones of the keto acid products and verified the presence of the predicted keto acid.

Henriksen and Closs (3) focused their efforts on the metabolism of phenylalanine by strains of *Proteus* and as early as 1938 described a rapid method for detecting the end product of the oxidative deamination of this amino acid—phenylpyruvic acid. A heavy suspension of the organism to be tested was made in a buffered saline solution containing L-phenylalanine. The presence of phenylpyruvic acid was detected using ferric ammonium sulfate which produced a green color with this product. Positive results could be obtained within 1 minute, but maximum color production required 1-hour incubation. Henriksen (4) made minor modifications of this method in 1950, including extending the incubation time to 4 hours.

In 1955 Singer and Volcani (5) substituted ferric chloride for ferric ammonium sulfate previously used to detect products of amino acid oxidative deamination. They suspended the organisms to be tested in saline and added an equal volume of L or DL amino acid in saline. The mixture was shaken for 15–30 minutes on a Kahn shaker at room temperature, followed by the addition of a few drops of ferric chloride. They studied a large number of the major genera in the family Enterobacteriaceae including the four species of *Proteus* and *Providence*. They found that *Proteus* and *Providence* deaminated histidine, leucine, isoleucine, norleucine, methionine, noraline, tryptophan, and phenylalanine. The addition of ferric chloride produced different colors depending on the alpha keto acid product. For example, the colors produced with ferric chloride for the corresponding keto acid product of histidine, leucine, isoleucine, norleucine, methionine, norvaline, and phenylalanine were green, grayish violet, orange, orange, violet, orange, and green, respectively. With the indole-3-pyruvic acid formed by the deamination of tryptophan, a cherry red color was formed when ferric chloride was added. Since the cherry red color was very stable these investigators (5) recommended that tryptophan deamination rather than phenylalanine deamination be used for routine testing (the green color produced with the ferric ion and phenylpyruvic acid faded rapidly). They also evaluated their positive findings using the ferric chloride test with hydrazone formation and found complete agreement. It is interesting to note that with alanine, arginine, aspartic acid, cysteine, cystine glutamic acid, glycine, lysine, ornithine, proline, hydroxyproline, serine, threonine, valine, and tyrosine, following incubation with the *Proteus* and *Providence* species, the color produced when ferric chloride was added was almost indistinguishable from the reagent itself.

Various formulations were developed for detecting the phenylalanine deaminase activity of *Proteus* and *Providence*. Shaw and Clark (6) devised a broth medium into which they incorporated malonate as well as phenylalanine. Ewing, Davis, and Reavis (7) evaluated this medium and recommended that, if a liquid medium was preferred, phenylalanine and malonate broths be prepared separately. Also a part of this study was an evaluation of phenylalanine agar. The production of phenylpyruvic acid was found to be easily detectable on phenylalanine agar after only 4 hours of incubation using ferric chloride; the green color of the reaction was more stable on agar. Following the work of Vassiliadis and Politi (8), in our laboratory we found (9) that the combination of urea and phenylalanine in a buffered yeast extract substrate with phenol red as an indicator provided a means for distinguishing between *Proteus* and *Providence* using a single tube. Urease-positive organisms produced a deep pink color in the medium. After reading the urease reaction in this system the medium was made slightly acid with dilute HCl; when a few drops of ferric chloride were added, phenylpyruvic acid could be detected.

When Edwards and Fife (10) developed lysine iron agar (LIA) in 1961, their main objective was to detect enteric pathogens such as *Salmonella* and *Arizona* which decarboxylate lysine; the main emphasis was the detection of *Arizona*. This medium is discussed more completely elsewhere in this book. The fact that *Proteus* and *Providence* could be distinguished from other members of the *Enterobacteriaceae* which were lysine decarboxylase negative was indeed a fortuitous finding and an addition to schemes for the identification of clinical isolates. *Proteus* and *Providence* produced red slants and acid butts on LIA. When the indicator was omitted from this medium a distinct orange color was produced throughout the slant. The red color that occurred with the indicator present was thought by these authors to be caused by *Proteus* and *Providence* deaminating lysine and producing an orange colored product which, when combined with the bromcresol purple, indicator, led to a distinctive red color.

To visualize the reactions involved in the deamination of phenylalanine and tryptophan, the structural formulas of the amino acid, the keto acid product, and the color complex formed by the product with ferric ion are shown below

$$\text{C}_6\text{H}_5\text{—CH}_2\text{—CHNH}_2\text{—COOH} \xrightarrow[\text{oxidase}]{\text{amino acid}}$$

Phenylalanine

$$\text{C}_6\text{H}_5\text{—CH}_2\text{—}\overset{\displaystyle \text{O}}{\overset{\displaystyle \|}{\text{C}}}\text{—COOH}$$

Phenylpyruvic acid

$+$

Fe^{3+}

$\downarrow$

blue-green color complex

$$\xrightarrow[\text{oxidase}]{\text{amino acid}}$$

Tryptophan

Indole pyruvic acid

$+$

Fe^{3+}

$\downarrow$

cherry red color complex

The exact nature of the color complexes formed by phenylpyruvic acid and indole pyruvic acid is not completely understood. Saifer and Harris (11) suggest that the reaction of phenylpyruvic acid with ferric ion may be similar to the ferric chloride thiosulfate reaction and might be written as

$$n\,(\text{PPA}^-) + m\,(\text{Fe}^{3+}) \;\rightleftarrows\; \underset{(\text{Fe}_m(\text{PPA})_n)}{\text{blue green}} \;\rightleftarrows\; m\,\text{Fe}^{2+} + \text{X}_\text{A}$$

$$\text{and } \text{X}_\text{A} \longrightarrow \text{X}_\text{B}$$

They describe X_A as a reversible oxidation product and X_B as the final product, which no longer reacts with the ferric ion. This postulate helps to explain why further addition of ferrous or ferric ions to this reaction mixture, once started, does not reproduce color after the original reaction has proceeded to completion. The color reactions of indole pyruvic acid with ferric ions may be similar though the stability of the cherry red color would indicate that less of a state of equilibrium is produced.

Although the deamination reaction of many of the amino acids, including phenylalanine and tryptophan, is well understood, this is not the case with lysine. The first product of lysine deamination is α-Keto-E-aminocaproic acid, but Meister (12) points out that this compound is spontaneously coverted to Δ'-piperideine-2-carboxylic acid. This reaction is shown below.

$$
\begin{array}{ccc}
\text{CH}_2\text{NH}_2 & \xrightarrow[\text{acid oxidase}]{\text{amino}} & \text{CH}_2\text{NH}_2 \xrightarrow[\text{neously}]{\text{sponta-}} \\
| & & | \\
\text{CH}_2 & & \text{CH}_2 \\
| & & | \\
\text{CH}_2 & & \text{CH}_2 \\
| & & | \\
\text{CH}_2 & & \text{CH}_2 \\
| & & | \\
\text{CHNH}_2 & & \text{C}=\text{O} \\
| & & | \\
\text{COOH} & & \text{COOH} \\
\text{Lysine} & & \alpha\text{-Keto-E-aminocaproic acid}
\end{array}
$$

Δ^1-Piperideine-2-carboxylic acid

Even with this information in mind it is difficult to explain the reaction of *Proteus* and *Providence* on LIA and to attempt to postulate which compound is producing the red color complex on the slant of this medium. This is a provocative question whose clarification requires further investigation.

At the present time the amino acid phenylalanine is most frequently used to determine the production of amino acid oxidase by microorganisms, playing a major role in the identification of Enterobacteriaceae (13). Although the cherry red color complex of indole pyruvic acid, the end product of the deamination of tryptophan, with the ferric ion was found to be more stable (5), the detection of the deamination of this amino acid has been used less extensively. One reason for this was the high cost of tryptophan; results with phenylalanine, which was less costly, were comparable (7). The differences in costs of these two amino acids have now narrowed, and tryptophan can now be considered for use in deaminase determination.

REFERENCES

1. Bernheim, F., Bernheim, M. L. C., and Webster, M. D. Oxidation of certain amino acids by "resting" *Bacillus Proteus. J. Biol. Chem.* **110:**165–172, 1935.

2. Stumpf, P. K., and Green, D. E. L-Amino acid oxidase of *Proteus vulgaris. J. Biol. Chem.* **153:**387–399, 1944.

3. Henriksen, S. D., and Closs, K. The production of phenylpyruvic acid by bacteria. *Acta. Pathol. Microbiol. Scand.* **15:**101–113, 1938.

4. Henriksen, S. D. A comparison of the phenylpyruvic acid reaction and urease test in the differentiation of *Proteus* from other enteric organisms. *J. Bact.* **60:**225–231, 1950.

5. Singer, J. and Volcani, B. E. An improved ferric chloride test for differentiating *Proteus-Providence* group from other Enterobacteriaceae. *J. Bact.* **69:**303–306, 1955.

6. Shaw, C., and Clarke, P. H. Biochemical classification of *Proteus* and *Providence* cultures. *J. Gen. Microbiol.* **13:**155–161, 1955.

7. Ewing, W. H., Davis, B. R. and Reavis, R. W. Phenylalanine and malonate media and their use in enteric bacteriology. *Publ. Health Lab.* **15:**153–167, 1957.

8. Vassiliadis, P., and Politi, G. Combined medium for the detection of urease production and L-phenylalanine deamination. *Ann. Inst. Pasteur.* **114:**431–435, 1968.

9. Ederer, G. M., Chu, J. H., and Blazevic, D. J. Rapid test for urease and phenylalanine deaminase production. *Appl. Microbiol.* **21:**545, 1971.

10. Edwards, P. R., and Fife, M. A. Lysine-iron agar in the detection of *Arizona* cultures. *Appl. Microbiol.* **9:**478–480, 1961.

11. Saifer, A., and Harris, A. F. Studies on the photometric determination of phenylpyruvic acid in urine. *Clin. Chem.* **5:**203–217, 1959.

12. Meister, A. *Biochemistry of Amino Acids*, Vol. 2, 2nd ed. Academic Press, New York, 1965, p. 944.

13. Edwards, P. R., and Ewing, W. H. *Identification of Enterobacteriaceae*, 3rd ed. Burgess Publishing Co., Minneapolis, 1972.

CHAPTER 8

DECARBOXYLASE TESTS

The ability of organisms to decarboxylate amino acids provides a means of differentiating their genera or species. The activity of a specific decarboxylase enzyme produced by an organism results in the breakdown of the amino acid and the formation of the corresponding amine with liberation of CO_2. The end products can be determined by manometric measurement of the CO_2 gas produced, by identification of the products produced using chromatographic procedures, or by an indicator system to demonstrate a shift in pH to alkaline.

There were reports of amine production resulting from bacterial putrefaction as early as 1908. It was not until 1946, however, when Gale (1) expanded and summarized the then existing knowledge of bacterial decarboxylase activity, that the diagnostic value of the detection of the presence of these enzymes became apparent. Gale described the specificity and reactions of the decarboxylase enzymes for six amino acids: lysine, ornithine, arginine, tyrosine, histidine, and glutamic acid. He stated that certain conditions must exist in order for active amino acid decarboxylase to be formed by bacteria. These included the following points: the organism must possess the potential to elaborate a specific decarboxylase enzyme, the growth must take place in a specific amino acid substrate, the organism must be able to synthesize a codecarboxylase enzyme or it must be added to the medium, the medium must be acid, the optimum temperature for decarboxylase production must be used for incubation, and decarboxylase enzymes will be formed only at the end of active cell division. The decarboxylase reactions for ornithine, lysine, arginine, and glutamic acid are listed below:

$$H_2N-\overset{\overset{\displaystyle H}{|}}{\underset{\underset{\displaystyle H}{|}}{C}}-\overset{\overset{\displaystyle H}{|}}{\underset{\underset{\displaystyle H}{|}}{C}}-\overset{\overset{\displaystyle H}{|}}{\underset{\underset{\displaystyle H}{|}}{C}}-\overset{\overset{\displaystyle H}{|}}{\underset{\underset{\displaystyle NH_2}{|}}{C}}-COOH \xrightarrow[\text{decarboxylase}]{\text{ornithine}} H_2N-\overset{\overset{\displaystyle H}{|}}{\underset{\underset{\displaystyle H}{|}}{C}}-\overset{\overset{\displaystyle H}{|}}{\underset{\underset{\displaystyle H}{|}}{C}}-\overset{\overset{\displaystyle H}{|}}{\underset{\underset{\displaystyle H}{|}}{C}}-\overset{\overset{\displaystyle H}{|}}{\underset{\underset{\displaystyle H}{|}}{C}}-NH_2 + CO_2$$

Ornithine   Putrescine

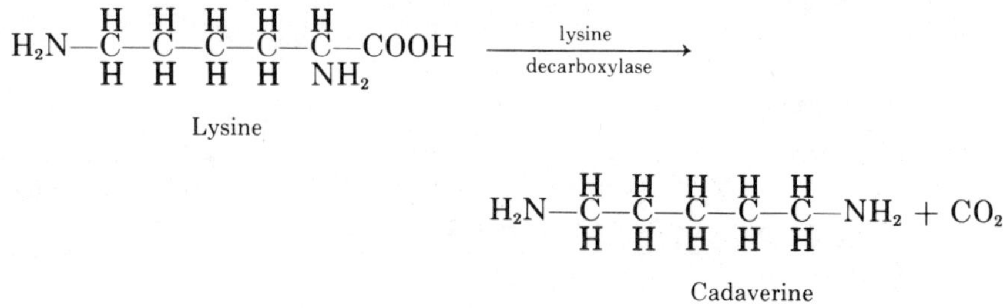

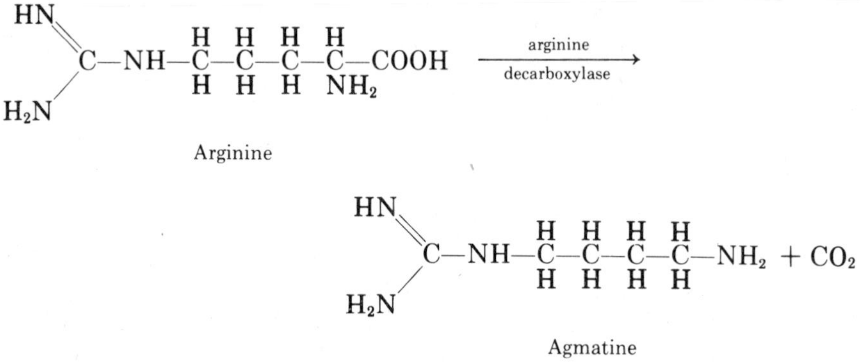

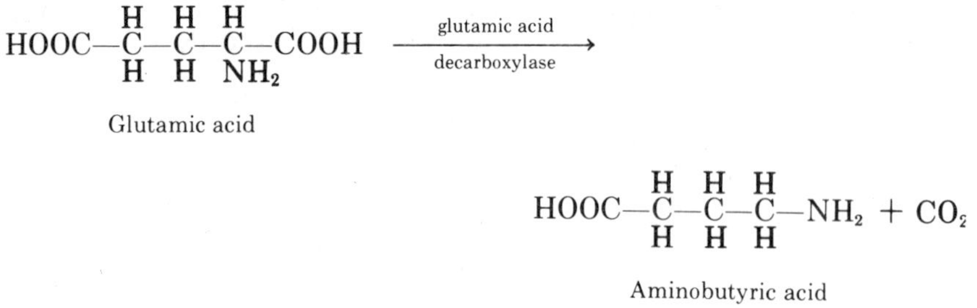

$$H_2N-\underset{\underset{H}{|}}{\overset{\overset{H}{|}}{C}}-\underset{\underset{H}{|}}{\overset{\overset{H}{|}}{C}}-\underset{\underset{H}{|}}{\overset{\overset{H}{|}}{C}}-\underset{\underset{H}{|}}{\overset{\overset{H}{|}}{C}}-\underset{\underset{NH_2}{|}}{\overset{\overset{H}{|}}{C}}-COOH \xrightarrow[\text{decarboxylase}]{\text{lysine}}$$

Lysine

$$H_2N-\underset{\underset{H}{|}}{\overset{\overset{H}{|}}{C}}-\underset{\underset{H}{|}}{\overset{\overset{H}{|}}{C}}-\underset{\underset{H}{|}}{\overset{\overset{H}{|}}{C}}-\underset{\underset{H}{|}}{\overset{\overset{H}{|}}{C}}-\underset{\underset{H}{|}}{\overset{\overset{H}{|}}{C}}-NH_2 + CO_2$$

Cadaverine

$$\underset{H_2N}{\overset{HN}{\diagdown\!\!\diagup}}C-NH-\underset{\underset{H}{|}}{\overset{\overset{H}{|}}{C}}-\underset{\underset{H}{|}}{\overset{\overset{H}{|}}{C}}-\underset{\underset{H}{|}}{\overset{\overset{H}{|}}{C}}-\underset{\underset{NH_2}{|}}{\overset{\overset{H}{|}}{C}}-COOH \xrightarrow[\text{decarboxylase}]{\text{arginine}}$$

Arginine

$$\underset{H_2N}{\overset{HN}{\diagdown\!\!\diagup}}C-NH-\underset{\underset{H}{|}}{\overset{\overset{H}{|}}{C}}-\underset{\underset{H}{|}}{\overset{\overset{H}{|}}{C}}-\underset{\underset{H}{|}}{\overset{\overset{H}{|}}{C}}-\underset{\underset{H}{|}}{\overset{\overset{H}{|}}{C}}-NH_2 + CO_2$$

Agmatine

$$HOOC-\underset{\underset{H}{|}}{\overset{\overset{H}{|}}{C}}-\underset{\underset{H}{|}}{\overset{\overset{H}{|}}{C}}-\underset{\underset{NH_2}{|}}{\overset{\overset{H}{|}}{C}}-COOH \xrightarrow[\text{decarboxylase}]{\text{glutamic acid}}$$

Glutamic acid

$$HOOC-\underset{\underset{H}{|}}{\overset{\overset{H}{|}}{C}}-\underset{\underset{H}{|}}{\overset{\overset{H}{|}}{C}}-\underset{\underset{H}{|}}{\overset{\overset{H}{|}}{C}}-NH_2 + CO_2$$

Aminobutyric acid

Gale studied the decarboxylation reaction manometrically by determining CO_2 evolution. In his work with living organisms using washed suspensions he found that the specific decarboxylase activity of a number of organisms was greatly enhanced at pH 5 or less and was completely depressed or negligible at pH 8.0 or greater. Since many organisms grew poorly under acid conditions, he added glucose to the medium which, when fermented, produced a suitable acid environment. He found 25°C to be the optimum temperature for decarboxylase production and determined that, since growth ceased at 16 hours, the enzymes could be harvested and purified at that time to prove their specificity. Gale found the decarboxylases to be adaptive enzymes and not produced in substrates free from amino acids, except for glutamic acid decarboxylase which can be formed in the absence of amino acids. When working with purified decarboxylase enzymes he found that the optimum pH for activity was often higher than the optimum pH for the intact cell. When decarboxylase enzymes were precipitated using ammonium sulfate at an alkaline pH the protein precipitate showed no activity toward the specific amino acid although activity was high before denaturation. Activity could be restored to the enzyme for a specific substrate by adding codecarboxylase in the form of a boiled preparation of lysine, arginine, ornithine, or tyrosine decarboxylase. The histidine and glutamic acid decarboxylase, however, differed from the other four decarboxylases studied by Gale in that they did not appear to possess codecarboxylase as a part of their enzyme systems. Codecarboxylase was also found in yeast or bacterial cells and in animal and plant tissues. A method for preparing codecarboxylase from dried yeast was devised by Gale and his associate. A synthetic product, pyridoxal phosphate, was compared with codecarboxylase from yeast and found to be effective in activating denatured decarboxylase enzymes for tyrosine, lysine, arginine, and ornithine.

During the 1950s Moeller became interested in the distribution of decarboxylases of lysine, arginine, ornithine, and glutamic acid among the Enterobacteriaceae with the thought that detection of these enzymes might aid in identifying organisms within this family more easily. In his early studies he determined decarboxylation manometrically but, unlike Gale, he used aqueous washed suspensions of organisms treated with toluene (2). The toluene apparently had a solvent effect on cell wall lipids and allowed for release of more enzyme. Although the determination of decarboxylases was helpful in identifying genera and species within the Enterobacteriaceae, the manometric methods were too cumbersome for routine use. To obviate this problem Moeller adapted his method of detecting decarboxylases in toluene-treated organisms from quantitating CO_2 production to determining a pH shift to alkaline when the corresponding alkaline amine was formed by the decarboxylation reaction (3). Using bromphenol red or bromcresol green as indicators, positive reactions could be detected within 1 minute with high activity; most positive reactions occurred within 25 minutes, and if no change in the indicator had taken place after 18 hours the reaction was considered negative. A paper chromatographic method was used to prove that the pH shift was due to amine formation. Encouraged by these results Moeller modified the method of determining pH shift in washed, toluene-treated bacteria to detection of decarboxylase enzymes of living bacteria in a specially formulated medium (4). The medium consisted of peptone, beef extract, pyridoxal, and glucose with two indicators—bromcresol purple and cresol red. The pH of the medium was adjusted to pH 6.0. A 1%

concentration of the L-amino acid to be investigated was incorporated in the medium. A control tube without amino acid was used with each organism tested.

In Moeller's early work on a medium to determine decarboxylase enzymes he found that even when the surface of the medium was kept small in comparison to total volume in the tube, the pH in the control tube, without amino acid, would become alkaline after the initial acid reaction occurred with the fermentation of glucose. Since the alkaline reaction appeared to start at the surface of the medium, he believed it to be due to oxidation. To overcome this problem he incubated the cultures anaerobically in an evacuated chamber. Later he found that a layer of sterile paraffin oil added to the surface of the medium was equally effective and made the procedure much more useful for diagnostic work. The test organisms in the amino acid medium and control tube were incubated at 37°C and observed daily. The control tube remained yellow because of the acid shift caused by fermentation of glucose, and tubes containing organisms that lacked specific decarboxylase for the added amino acid also remained yellow. The color of the medium changed to violet when a specific decarboxylase was present and indicated that the alkaline amine of the acid had been formed. Moeller felt that those reactions that occurred in the first 24–48 hours of incubation appeared to have diagnostic value in the differentiation of *Salmonella*, *Escherichia coli*, *Klebsiella*, and other organisms belonging to Enterobacteriaceae. When the incubation was extended to 6–10 days, weak reactions were seen that he considered to be due to weak decarboxylase activity as well as slow oxidations. The decarboxylase method of Moeller has become the reference method for decarboxylase tests, using an incubation time of 4 days.

Investigators continued to develop decarboxylase test methods that would be useful in the diagnostic laboratory. Carlquist (5) in 1956 developed a simple medium containing casitone, dextrose, and lysine without indicators and adjusted to pH 7.5. Organisms were incubated in this medium at 37°C for 18–24 hours; strong NaOH was then added, followed by the addition of chloroform and vigorous shaking. The specimen was centrifuged and the chloroform layer removed. The lysine decarboxylation product cadaverine, which is soluble in chloroform, was detected by adding ninhydrin. Positive reactions were indicated when a purple color was formed. Centrifugation and extraction were necessary to separate cadaverine from lysine, which also gives a purple color with ninhydrin.

Falkow (6) developed another basal medium in 1958 consisting of peptone, glucose, yeast extract, and a single indicator, bromcresol purple, adjusted to pH 6.7–6.8, and 0.5% concentration of L-lysine. After 24 hours a yellow color indicated a negative reaction and a purple color showed a positive reaction. In this original method neither a control tube without amino acid nor a paraffin oil overlay was recommended. Ewing, Davis, Edwards (7) evaluated this decarboxylase medium with lysine, ornithine, and arginine, comparing it with Moeller's medium using a paraffin oil seal and control tube for each test for both procedures. Their evaluation showed that there was little difference between the results with Moeller's or Falkow's medium when *Shigella, Salmonella, Arizona, Citrobacter, Serratia,* and *Hafnia* groups were studied. However, differences did occur with *Klebsiella, Aerobacter (Enterobacter), Proteus,* and *Escherichia.* There were no important differences for these organisms at the 24-hour reading, but when the cultures were reincubated the indicator in the Falkow medium became decolorized, or the reaction in the control tube became alkaline making it impossible to interpret the results. Ewing and his colleagues, on the basis of these findings, recommended that a control tube without amino acid should always be included with each test and that paraffin oil seals should be used on all decarboxylase test media. They also recommended that Moeller's decarboxylase medium be used as a reference method.

In the preceding work it appeared that 24 hours was sufficient incubation time to determine lysine decarboxylation. Making use of this knowledge Edwards and Fife (8) developed a solid medium for the lysine decarboxylase test. Agar was incorporated into medium containing peptone, yeast extract, glucose, L-lysine, ferric ammonium citrate, sodium thiosulfate, and bromcresol purple for an indicator. After sterilization the tube was slanted and the organism was inoculated on the slant and stabbed to the butt. After 24 hours negative reactions showed a purple slant with a yellow butt and in positive reactions the slant and butt were purple when alkaline cadaverine was formed. The hydrogen sulfide indicator present in the medium proved to be an additional diagnostic acid. With this medium it was also possible to detect deamination reactions in which the slant was red and the butt yellow.

The determination of ornithine decarboxylase activity was incorporated with tests for indole production and assessment of motility in 1970 (9). This medium was similar to Falkow's except for the addition of a small (0.2%) amount of agar. The agar provided sufficient anaerobiosis in the butt of the medium to dispense with overlaying the medium with paraffin oil to determine ornithine decarboxylation. Negative ornithine reactions showed yellow butts with a rim of purple at the top of the tube caused by oxidation; the butt was purple in positive reactions with a darker rim of purple at the top of the tube. Reactions in this medium are comparable to Falkow's and Moeller's media.

The detection of decarboxylase activity in nonfermentative bacteria has not been simple. The use of pH indicators, as in the work previously mentioned, is not definitive if glucose has not been fermented with an indicator change in color denoting acid conditions. It is possible to determine decarboxylase activity in nonfermentative bacteria by careful comparison of the control tube without amino acid, with the tube containing a specific amino acid. The pH of the resultant amine in the medium will be higher than the pH of the control, and the indicator will have a deeper color. Since an indicator system is not used in the Carlquist method (5) this procedure is especially useful for determining decarboxylase reactions in nonfermenters.

Thin-layer chromatography (TLC) has proved to be a useful, definitive, and rapid means (3–4 hours) of determining arginine decarboxylase activity occurring in conjuction with arginine dihydrolase activity in *Pseudomonas maltophilia* (10,11). In our laboratory (11) we used a heavy suspension of the organism in 0.25 ml of a 1% L-arginine hydrochloride aqueous solution at pH 6.4. After incubation for 2.5 hours at 37°C the supernate was applied to a cellulose precoated sheet along with arginine, ornithine, citrulline, and agmatine standards. After development in *n*-butanol-acetone-acetic acid-water (35:35:10:20) the chromatogram was removed, dried, sprayed with ninhydrin, and again dried on a slide drier for development of the purple color formed by ninhydrin reacting with amino acids and amines. Using this method the various amino acids and amines migrated different distances up the chromatogram, allowing for their identification by comparison with the standards. When indicator systems are used to detect arginine decarboxylase activity, negative results are obtained, since the reaction does not cause a shift to alkaline pH. The use of TLC, therefore, represents a simple and definitive technique to determine specific end products of enzymatic reactions. As more investigation is done using TLC, the methods will undoubtedly continue to be improved and simplified. Already Zolg and Ottow (12) have shown that by choosing more effective solvents, butanol-acetone-diethylamine-water (10:10:2:5), for TLC the method we have described can be improved.

Fay and Barry (13) described a rapid ornithine decarboxylase test using a medium similar to Falkow's. The pH of the medium was adjusted to 5.5 and dispensed in 0.5-ml amounts. Single colonies were emulsified in the medium, and the mixture was overlaid with sterile mineral oil. Ornithine decarboxylase activity was detected in 2–4 hours by noting the development of a deep purple color due to the alkaline reaction. The pH of the negative tests decreased from pH 5.5 to 5.2–5.0 and gave a definite yellow color. In our laboratory (14) we modified this medium for a rapid 4-hour lysine decarboxylase test by substituting bromthymol blue for bromcresol purple. The pH endpoint for the rapid lysine decarboxylase test is 6.1 just as the yellow-to-green color breakpoint occurs with bromthymol blue. By changing the indicator to bromthymol blue there was 100% correlation with reference methods.

Today tests for decarboxylase activity are used extensively in the diagnostic microbiology laboratory to aid in the identification of organisms within the Enterobacteriaceae (15). It must be kept in mind when performing decarboxylase tests using pH indicators and substrates that contain peptones that air must be excluded by means of a paraffin oil seal or by making the medium solid or semisolid with agar. These measures will prevent alkaline oxidation of the medium which would lead to erroneous results. The use of a screw cap tube does not suffice. A control tube without amino acid should always be used in conjunction with the medium containing the specific amino acid. Air need not be excluded when the products are extracted using differential solvents and their presence confirmed by reacting with ninhydrin. Likewise, it is not necessary to exclude air when evidence of decarboxylation is detected by TLC using substrates containing specific amino acids without nutrients.

REFERENCES

1. Gale, E. F. The bacterial amino acid decarboxylases. In Nord, F. F,, Ed. *Advances in Enzymology and Related Subjects of Biochemistry*, Vol. 6. Interscience Publishers, New York, 1946.

2. Moeller, V. Activity determination of amino acid decarboxylases in Enterobacteriaceae. *Acta. Pathol. Microbiol. Scand.* **34**:102–114, 1954.

3. Moeller, V. Distribution of amino acid decarboxylases in Enterobacteriaceae. *Acta. Pathol. Microbiol. Scand.* **35**:259–277, 1954.

4. Moeller, V. Simplified tests for some amino acid decarboxylases and for the arginine dihydrolase system. *Acta. Pathol. Microbiol. Scand.* **36**:158–172, 1955.

5. Carlquist, P. R. A biochemical test for separating paracolon groups. *J. Bact.* **71**:339–341, 1956.

6. Falkow, S. Activity of lysine decarboxylase as an aid in the identification of *Salmonellae* and *Shigellae*. *Am. J. Clin. Pathol.* **29**:598–600, 1958.

7. Ewing, W. H., Dabis, B. R., and Edwards, P. R. The decarboxylase reactions of Enterobacteriaceae and their value in taxonomy. *Publ. Health. Lab.* **18**:77, 1960.

8. Edwards, P. R. and Fife, M. A. Lysine iron agar in the detection of *Arizona* cultures. *Appl. Microbiol.* **9**:478–480, 1961.

9. Ederer, G. M. and Clark, M. Motility indole ornithine medium. *Appl. Microbiol.* **20**:849–850, 1970.

10. Goldschmidt, M. C. and Lockhart, B. M. Rapid method for determining decarboxylase activity: Arginine decarboxylase. *Appl. Microbiol.* **22**:350–357, 1971.

11. Williams, G. A., Blazevic, D. J. and Ederer, G. M. Detection of arginine dihydrolase in nonfermentation gram-negative bacteria by use of thin-layer chromatography. *Appl. Microbiol.* **22:**1135–1137, 1971.

12. Zolg, W. and Ottow, J. C. G. Improved thin-layer technique for detection of arginine dihydrolase among *Pseudomonas* species. *Appl. Microbiol.* **26:**1001–1003, 1973.

13. Fay, G. D. and Barry, A. L. Rapid ornithine decarboxylase test for the identification of Enterobacteriaceae. *Appl. Microbiol.* **23:**710–713, 1972.

14. Brooker, D. C., Lund, M. E., and Blazevic, D. J. Rapid test for lysine decarboxylase activity in Enterobacteriaceae. *Appl. Microbiol.* **26:**622–623, 1973.

15. Edwards, P. R. and Ewing, W. H. *Identification of Enterobacteriaceae,* 3rd ed. Burgess Publishing Co., Minneapolis, 1972.

CHAPTER 9

DEOXYRIBONUCLEASE (DNase) TEST

The ability of an organism to break down deoxyribonucleic acid (DNA) by means of a deoxyribonuclease has been used clinically to differentiate between groups of microorganisms (1, 2, 3), to aid in determining the potential pathogenicity of staphylococci (4), and to determine the immunological response to group A streptococcal infections (5).

DNA is a polynucleotide that is composed of a chain of deoxyribonucleotide units. Deoxyribonucleotides are phosphoric acid esters of nucleosides (a pyrimidine or purine base attached to deoxyribose). The phosphoric acid is attached to the deoxyribose ring at either the 3′ or 5′ position. Examples of deoxyribonucleotides are adenosine-5′-monophosphate (5′-AMP) and guanosine-5′-monophosphate (5′-GMP).

DNase is an extracellular enzyme that hydrolyzes DNA to yield oligonucleotides (chains of several deoxyribonucleotides). The DNase of *Serratia marcescens* yields di-, tri-, and tetranucleotides (6). The DNase of Group A streptococci produces predominantly products having a chain length of 3 to 10 deoxyribonucleotide units but also produces mononucleotides (5).

DNA has physical and chemical properties different from those of oligonucleotides or mononucleotides, and advantage has been taken of these differences to detect the hydrolysis of DNA by DNase. Not all of the many methods proposed to determine hydrolysis of DNA are applicable in the clinical laboratory. The principles of those most commonly used are presented here.

Probably the first method proposed for clinical use is that of Jeffries, Holtman, and Guse (7). This method is based on the fact that DNA is precipitated by acid, whereas the oligonucleotides liberated by hydrolysis are soluble in acid. The organisms are streaked on a plate of a nutrient medium (DNase Test Agar) containing DNA; after incubation the plate is flooded with 1 *N* HCl. Unhydrolyzed DNA is precipitated, causing the medium to become cloudy. If the organism being tested produces DNase there will be a clear zone around the organism, indicating the hydrolysis of DNA. Determination of DNase activity based on this principle may also be carried out on broth cultures utilizing a medium containing DNA (8). Because of the turbidity of the culture itself, the cells must first be separated from the fluid by centrifugation (the extracellular DNase is released into the culture fluid during growth). Then HCl is added to the supernatant fluid; the turbidity which develops is compared on a spectrophotometer with the turbidity of an uninoculated medium. A decrease in turbidity in the culture fluid indicates the hydrolysis of DNA. This method lends itself well to quantitation of DNase activity.

In 1969 Schreier (9) modified DNase test medium by the addition of Toluidine Blue O. Toluidine blue is a metachromatic dye. Metachromasia is the property whereby a dye will not stain true because of complexes formed with some substances which result in an absorption spectrum different from the original dye (10). The true, or orthochromatic, staining of toluidine blue is blue, whereas metachromatic staining results in a pink to red to violet color (11). When toluidine blue is complexed with DNA it stains blue, but when complexed with oligonucleotides or mononucleotides the resulting absorption spectrum results in metachromatic (or pink) staining. Thus when organisms are grown on DNase agar with toluidine blue those strains that produce DNase will have a pink zone around the colonies. The advantage of this method over flooding the plate with HCl is that no addition of reagents is required, the plates can be observed for short or long periods of incubation, and the organisms can be subcultured from the plate after reading the test. However, the toludine blue can inhibit gram-positive bacteria, so that staphylococci may not grow on this medium.

Another method utilizing toluidine blue as an indicator was described by Lachica et al. (12), and is called the metachromatic agar-diffusion microslide technique. In this method a buffered agar containing DNA and toluidine blue is prepared on plastic immunoplates. After hardening, 2-mm wells are cut in the plate. The wells may be filled with broth cultures or suspensions of the organism to be tested and then incubated at 37°C. After 3 hours of incubation the tests may be read; development of a pink zone around the well indicates production of deoxyribonuclease. These workers found that the method could also be used as a quantitative determination of DNase activity on the basis of the diameter of the pink zone developed. This method also has the advantage of being rapid and can be used with gram-positive organisms.

An improved medium for detecting DNase production was developed that has the advantages of the toludine blue medium but not its disadvantage (13).

This method utilized DNase test medium to which is added methyl green. Methyl green combines with DNA to form a green complex at pH 7.5. If the DNA is hydrolyzed, the methyl green is released and becomes a colorless compound (14). Therefore, the uninoculated medium with DNA is green; if an organism produces DNase a colorless zone will be seen around the colonies. Gram-positive organisms will grow well on this medium.

A rapid test for detection of bacterial DNase was described by Wolf et al in 1969 (15). This test is an indirect determination of DNase activity because it does not use DNA as a substrate. Rather, the substrate in the method is 5-bromo-4-chloro-3-indolyl-thymidine-3′-phosphate, a substance with phosphate bonds similar to those in DNA which are cleaved by DNase. Upon hydrolysis, the colorless indolyl substrate is converted to insoluble blue-green indigo. The reaction occurs as follows:

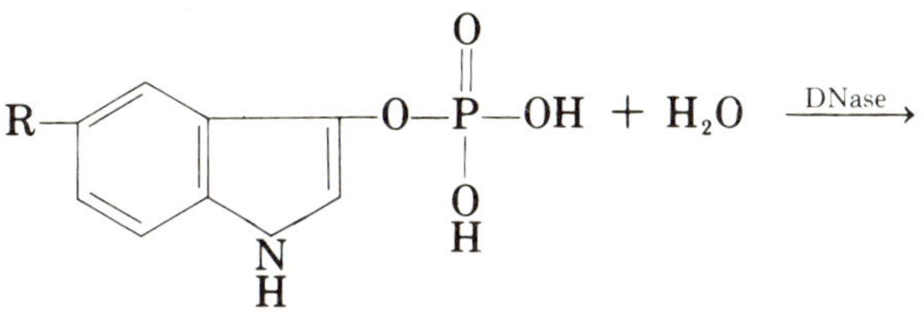

Indolyl phosphate

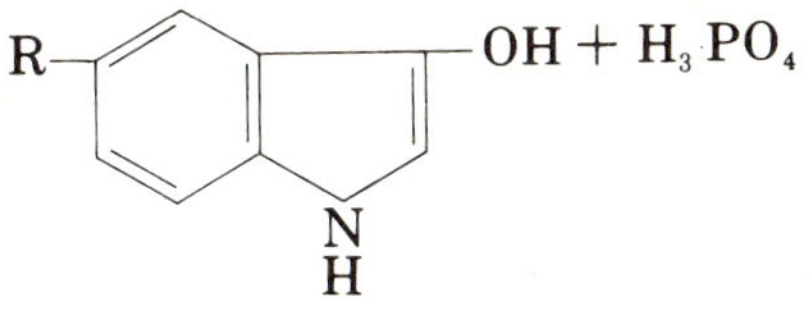

Indolyl

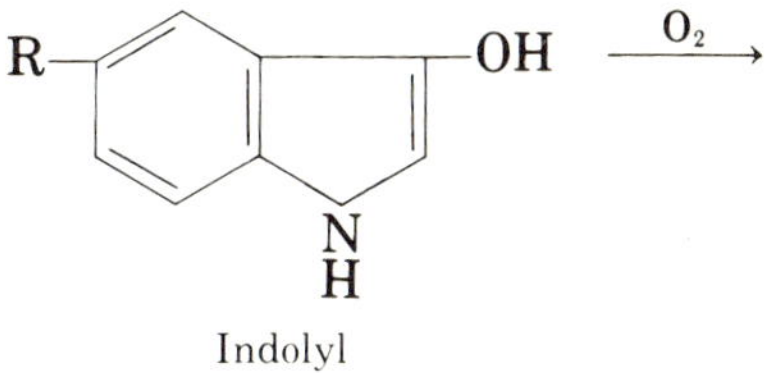

Indolyl

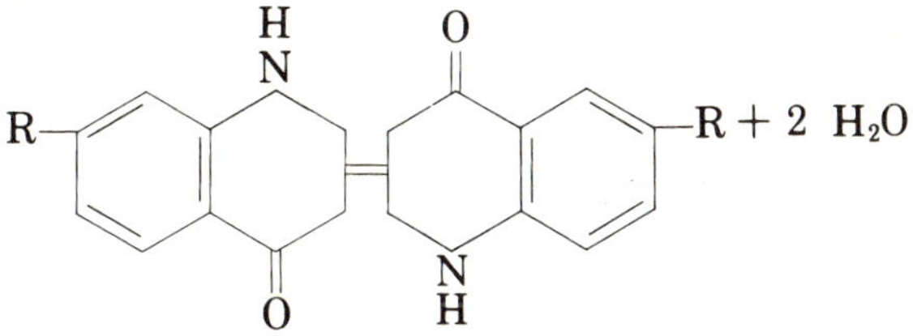

Blue-green indigo

The reaction occurs rapidly, and a positive test (blue-green color) may be detected within 4 hours. The substrate is very expensive, however.

Most of the bacterial DNases studied have been found to require the presence of divalent cations for their activity (5, 6). Media containing peptones will usually contain these cations. The pH optima have also been investigated and have been found to range from 5.5 to 8.5 (5, 16). For clinical purposes, media with a pH of 7.2 have been found to be satisfactory.

REFERENCES

1. Rothberg, N. W. and Swartz, M. N. Extracellular deoxyribonucleases in members of the family *Enterobacteriaceae. J. Bact.* **90**:294–295, 1965.

2. Blazevic, D. J. Identification of *Serratia* in the diagnostic microbiology laboratory. *Am. J. Clin. Pathol.* **51**:277–279, 1969.

3. Martin, W. J. and Ewing, W. H. The deoxyribonuclease test as applied to certain gram-negative bacteria. *Canad. J. Microbiol.* **13**:616–618, 1967.

4. Weckman, B. G. and Catlin, B. W. Deoxyribonuclease activity of micrococci from clinical sources. *J. Bact.* **73**:747–753, 1957.

5. Wannamaker, L. W. Streptococcal deoxyribonucleases. In Uhr, J. W., Ed. *The Streptococcus, Rheumatic Fever, and Glomerulonephritis.* Williams & Wilkins, Baltimore, 1964.

6. Nestle, M. and Roberts, W. K. An extracellular nuclease from *Serratia marcescens.* II. Specificity of the enzyme. *J. Biol. Chem.* **244**:5219–5225, 1969.

7. Jeffries, C. D., Holtman, D. F., and Guse, D. G. Rapid method for determining the activity of microorganisms on nucleic acids. *J. Bact.* **73**:590–591, 1957.

8. Erickson, A. and Deibel, R. H. Turbidimetric assay of staphylococcal nuclease. *Appl. Microbiol.* **25**:337–341, 1973:

9. Schreier, J. B. Modification of deoxyribonuclease test medium for rapid identification of *Serratia marcescens. Am. J. Clin. Pathol.* **51**:711–716, 1969.

10. Davidsohn, I. and Henry, J. B. Todd-Sanford Clinical Diagnosis by Laboratory Methods, 14th ed. W. B. Saunders Co., Philadelphia, 1969, p. 94.

11. Lillie, R. D. *J. H. Conn's Biological Stains*, 8th ed. Williams & Wilkins, Baltimore, 1969, p. 414.

12. Lachica, R. V. F., Hoeprich, P. D., and Franti, C. E. Convenient assay for staphylococcal nuclease by the metachromatic well-agar-diffusion technique. *Appl. Microbiol.* **24**:920–923, 1972.

13. Smith, P. B., Hancock, G. A. and Rhoden, D. L. Improved medium for detecting deoxyribonuclease-producing bacteria. *Appl. Microbiol.* **18**:991–993, 1969.

14. Kurnick, N. B. The determination of deoxyribonuclease activity by methyl green; application to serum. *Arch. Biochem.* **29**:41–53, 1950.

15. Wolf, P. L., Horwitz, J., Mandeville, R., Vazquez, J., and von der Muehll, E. A new and unique method for detecting bacterial deoxyribonuclease in the clinical laboratory. *Tech. Bull. Reg. Med. Tech.* **39**:83–86, 1969.

16. Nestle, M. and Roberts, W. K. An extracellular nuclease from *Serratia marcescens.* I. Purification and some properties of the enzyme. *J. Biol. Chem.* **244**:5213–5218, 1969.

CHAPTER 10

ESCULIN HYDROLYSIS

The ability of organisms to hydrolyze esculin is used chiefly to differentiate group D streptococci from other groups of streptococci. Esculin hydrolysis has also been used to characterize gram-negative rods and anaerobic bacteria.

The value of determining the ability of organisms to hydrolyze esculin (aesculin) was first observed by Harrison and van der Leck (1) in 1909 while evaluating bacteriological methods for determining fecal contamination of water. Using a medium composed of peptone, sodium taurocholate, esculin, ferric citrate, agar, and water they found that organisms now known as *Escherichia coli* and *Enterobacter* species hydrolyzed esculin and could be detected easily even when present in small numbers mixed with other flora. They proposed that bacteria with the ability to hydrolyze the glucoside, esculin, produced the following reaction in their newly devised medium:

$$C_{15}O_{16}O_9 + H_2O \quad \rightarrow \quad C_6H_{12}O_6 + C_9H_6O_4$$

Esculin   Glucose   Esculetin

+

ferric citrate

↓

dark brown salt

In 1924 Rochaix (2) showed that enterococci could be differentiated from other streptococci on the basis of esculin hydrolysis. Meyer and Schonfeld (3) tested a large number of enterococci and other streptococci using the medium of Harrison and van der Leck (1) and concluded that esculin hydrolysis was the best single test available to differentiate these two groups of streptococci. This diagnostic test for enterococci apparently fell somewhat into disrepute following the report of Weatherall and Dible (4) in 1929. Two points were made by these investigators: (*a*) the specificity of the test depended on the presence of bile salts which inhibited streptococci other than enterococci, or even more forcefully stated, esculin hydrolysis was only an indicator system for bile tolerance; and (*b*) the ability to split esculin was a specific characteristic of enterococci. Unfortunately, the last point was stated briefly in the conclusion whereas the first point was made in detail in the body of the paper; hence interest in this test lagged. Renewed interest in using esculin hydrolysis as a diagnostic characteristic for the identification of group D streptococci was brought about by the report of Swan (5) in 1954 and has continued in more recent times (6, 7, 8).

Although considerable work has been done to evaluate the usefulness of esculin hydrolysis as a diagnostic test, little new light has been shed on the biochemical nature of the reaction. In particular the composition of the dark brown salt has not been defined exactly.

The Haworth projection formulas for sugars (shown below) does provide some additional insight into the reaction. Also shown is a theoretical formula for the iron complex formed with esculetin (9).

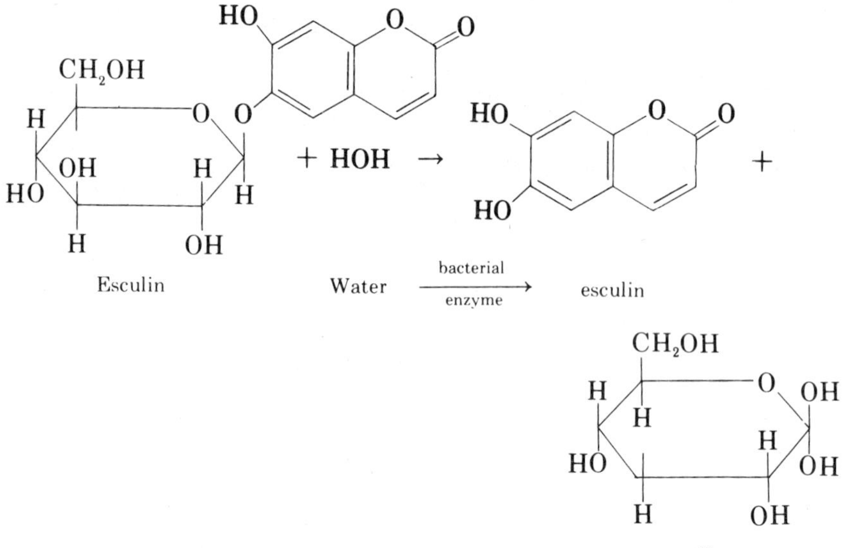

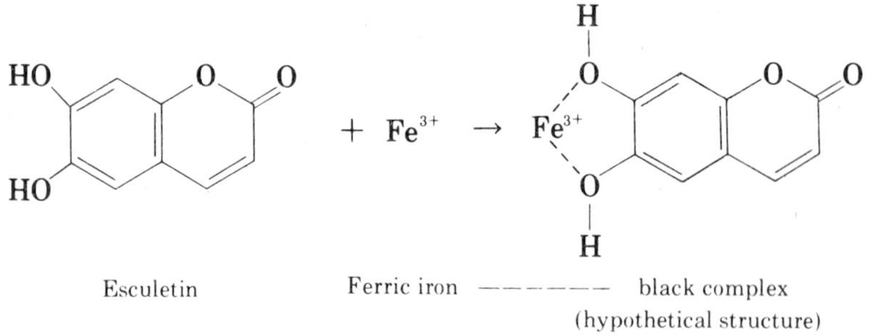

Esculin + Water $\xrightarrow[\text{enzyme}]{\text{bacterial}}$ esculin + β-D-Glucose

Esculetin + Fe^{3+} → black complex (hypothetical structure)

When using esculin hydrolysis as a diagnostic characteristic for organism identification, it is important to ascertain the complete contents of the medium used for the test. If any quantity of bile is present, negative tests must be evaluated, since the organism may be inhibited by bile. When the organism is sensitive to bile, adaptations of the esculin hydrolysis method of Gemmell and Hodgkiss (10) can be used. Esculin may be added to any basal medium in 1.0% concentration. Hydrolysis of esculin can be determined either by observing the loss of fluorescence in the medium under ultraviolet light or by detecting the appearance of precipitating coral-like white crystals assumed to be esculetin, or by the black color formed with ferric ion. Tests for esculin hydrolysis performed in esculin agar may not agree with results of this test performed in esculin broth. When working with anaerobes, esculin broth is recommended and the indicator, ferric ammonium citrate, should be added after incubation to detect hydrolysis (11).

REFERENCES

1. Harrison, F. C. and Van der Leck, J. Aesculin bile salt media for water analysis. *Bakt. Abt. 2* **22**:547, 1909.

2. Rochaix, A. Mileaux a l'esculine pour le diagnostic differential des bacteries due groupe Strepto-Entero Pneumocoque. *Ct. R. Soc. Biol.* **90**:771–772, 1924.

3. Meyer, V. K. and Schonfeld, H. Über die Unterscheidung des Enterococcus vom Streptococcus viridans und die Beziehungen beider zum Streptococcus lactis. *Zentralbl. Bakt. Abt. 1 orig.* **99**:402–418, 1926.

4. Weatherall, C. and Dible, J. H. Aesculin fermentation and hemolysis by enterococci. *J. Pathol. Bact.* **32**:413–417, 1929.

5. Swan, A. The use of bile-aesculin medium and of Maxted's technique of Lancefield grouping in the identification of enterococci, group D streptococci. *J. Clin. Pathol.* **7**:160–163, 1954.

6. Sabbaj, J. Sutter, V. L., and Finegold, S. M. Comparison of selective media for isolation of presumptive group D streptococci from human feces. *Appl. Microbiol.* **22**:1008–1011, 1971.

7. Facklam, R. R. Recognition of group D streptococcal species of human origin by biochemical and physiological tests. *Appl. Microbiol.* **23**:1131–1139, 1972.

8. Lee, W. Improved procedure for the identification of group D enterococci with two new media. *Appl. Microbiol.* **24**:1–3, 1972.

9. *The PathoTec Rapid I-D System.* General Diagnostics Division of Warner-Lambert, Morris Plains, N.J., 1973, p. 12.

10. Gemmell, M. and Hodgkiss, W. The physiological characters and flagellar arrangement of motile homofermentative *Lactobacilli. J. Gen. Microbiol.* **35**:519–526, 1964.

11. Holdeman, L. V. and Moore, W. E. C. *Anaerobe Laboratory Manual.* V. P. I. Anaerobe Laboratory, Virginia Polytechnic Institute and State University, Blacksburg, Va., 1972.

CHAPTER 11

GELATIN LIQUEFACTION

The ability of an organism to hydrolyze, or liquefy, gelatin has been very useful in the differentiation of many genera and species of bacteria, including those in the Enterobacteriaceae, the Pseudomonadaceae, and the anaerobes, particularly the *Clostridium* genus.

Gelatin is a protein that is derived from collagen. Collagen (one of the principal components of connective tissue) is an insoluble protein, but can be readily converted into the soluble protein gelatin by boiling. The chemical structures of gelatin and collagen are similar; it is only the physical organization that is different (1). These proteins have an unusually high content of the amino acids, proline, and glycine content.

Gelatin is hydrolyzed by the gelatinase of various bacteria. In some organisms, such as *Clostridium perfringens* (*welchii*), the enzyme is called a collagenase (2). Upon hydrolysis into its constituent amino acids, gelatin loses its gelling properties and remains a liquid, even at low (refrigerator) temperatures. Therefore, many tests for gelatin liquefaction are based on the concept of conversion of gelled gelatin to a liquid. In these tests the organism is generally inoculated or stabbed into a medium containing nutrients (e.g., peptone) and gelatin (3). Because many bacteria have been found to have greater gelatinase activity at temperatures lower than 37°C (4, 5), it is recommended that tests for gelatinase be incubated at 20–26°C for up to 30 days. This is especially important if one wishes to determine differences in rate of liquefaction. However, with organisms that possess high gelatinase activity (e.g., *C. perfringens* and *Serratia marcescens*) the test may be incubated at 37°C. In this case, since gelatin is liquid at this high temperature, the medium must be placed at a lower temperature after incubation to see if liquefaction has occurred. An uninoculated control tube should be treated in the same manner as the test.

The stab method for gelatin liquefaction has been in use for many years. However, the reproducibility of results has not always been good. In 1923

Levine and Carpenter (6) made a detailed study of gelatin liquefaction by bacteria. They found that the rate of liquefaction was dependent on the age of the gelatin medium because the viscosity of autoclaved gelatin increased upon storage at 22°C. They studied the action of seven organisms (genus and species not given) in a gelatin medium by following a decrease in viscosity and by formal titration.(Formal titration is a measure of the hydrolysis of proteins by detecting the formation of free amino acids as the peptide bonds of the protein are being broken down.) On this basis they classified the organisms into three groups:

1.   Nonliquefiers: gelatin not liquefied, no hydrolysis.
2.   Weak liquefiers: gelatin partially liquefied, little if any decomposition of the gelatin.
3.   Strong liquefiers: gelatin liquefied, decomposition of the gelatin noted.

On the basis of their studies Levine and Carpenter concluded that determining changes in viscosity by a viscosimeter is preferable to looking for liquefaction. Although not always practical in a clinical situation, their recommendation was one of the first attempts at standardizing a rather crude test.

Because of the unreliability of the gelatin stab method, many other investigators made further attempts to improve the detection of gelatin liquefaction. Frazier (7) developed a method based on the detection of a change in the composition of gelatin rather than its liquefaction. The gelatin was added to a buffered, nutrient agar medium in plates. The plates were inoculated heavily in the center, and incubation was recommended at 30°C for 48 hours. After incubation, the plate was flooded with an acid solution of mercuric chloride. Mercuric chloride causes precipitation of the gelatin in the medium; if the gelatin was decomposed by the test organism a clear area would be seen around the growth. This method was compared to the "standard" gelatin stab procedure and was found to be more reproducible and more sensitive. Interestingly, some organisms that failed to liquefy gelatin were found to cause a change in its composition as measured by the mercuric chloride test, indicating the presence of a gelatinase, perhaps either of low activity or present in only small quantities. The advantages of Frazier's method over the stab method are its sensitivity, shorter time of incubation, and lack of interference in the test if older gelatin is used. The disadvantage would be its inability to detect rate of liquefaction, which is considered to be important in some cases, such as differentiation within the tribe *Klebsielleae* (3).

A method based on the same principle as Frazier's was described in which the gelatin-agar medium was incorporated into a small glass column (2). This method was later modified by Clarke and Cowan (8) who used small Durham tubes. After inoculation and incubation the acidic mercuric chloride was added, and a clear zone under the meniscus indicated liquefaction. With this method positive results could be obtained in as little as 4 hours by using a heavy suspension of fresh growth from any suitable medium.

In 1953, Kohn reported on a new gelatin liquefaction method (9). This method used formalin-denatured gelatin combined with finely powdered charcoal. The gelatin binds charcoal particles together, and disks of the combination are used. Because of the formalin denaturization, the gelatin in the disks does not melt at 37°C. The disks are added to a suitable broth medium, which is then inoculated with the test organism; or the disk can be added to an already growing culture. If the organism liquefies the gelatin, the charcoal is released from the disk, and the medium becomes black throughout. Lautrop (4) modified Kohn's method somewhat by adding the disk to a heavy suspension of organisms prepared from a solid medium. This had the advantage of giving more rapid results because the enzymes were preformed (growth does not take place in this test), and a higher concentration of enzyme was present in the test. A rough indication of rate of liquefaction can be ascertained in this test, and positive tests can be seen in as little as 15 minutes with a very heavy suspension of a strongly liquefying organism. The test itself should be incubated at 37°C, although for optimal gelatinase production, when testing a wide variety of organisms, Lautrop recommended that the organisms be grown at 30°C for preparation of the inoculum.

During this study, Lautrop found that some bacteria (e.g., *Salmonella, Serratia, Staphylococcus*) produce a calcium-requiring gelatinase; he recommends that 0.01 M $CaCl_2$ be incorporated into the suspending medium. He also noted that the addition of toluene to the suspension seemed to increase the rapidity of the reaction, presumably by increasing the permeability of the cell wall and allowing for better release of intracellular enzymes. The effect of toluene was especially notable with organisms that were slow liquefiers. However, since the gelatinase of some organisms was inhibited by toluene, it is recommended that two tests be run—one with and one without toluene. Overall, the modified Kohn's test is at least as rapid (and often more rapid) than Frazier's test and has the advantage that no reagents have to be added. Thus continuous observation can be made leading to semiquantitation of the liquefaction rate.

Another approach to the rapid detection of gelatin liquefaction was first described by Pickford and Dorris in 1934 (10). They used the gelatin of an unexposed photographic plate as a substrate for the test; digestion of the gelatin was indicated by a clearing of the film within 2 hours. LeMinor and Piechaud

(11) developed a test based on a similar principle, but they used strips of exposed, developed film placed in a heavy suspension of organisms. Upon digestion of the gelatin the suspension becomes greyish black because of the release of the silver sulphide present in the gelatin film. Strong liquefiers gave positive results in 10 minutes to 6 hours. Weak liquefiers required up to 7 days to give a positive test. More rapid results were obtained if the tests were placed on a shaker.

Still another rapid method for gelatin liquefaction utilizes strips of exposed, undeveloped x-ray film which are placed into a heavy suspension of the organism to be tested (12–14). If the organisms liquefy gelatin the green gelatin emulsion of the film will be removed, leaving the clear, bluish film support. If very small strips are placed in a very small amount of a heavy organism suspension in a 6-mm tube, the rapidity of the test is increased (14).

It is obvious that no single test among the many that have been developed to detect liquefaction of gelatin is completely satisfactory. The results of all, it must be stressed, are not completely comparable, which must be taken into account when using the test for identification of previously described species.

REFERENCES

1. West, E. S. and Todd, W. R. *Textbook of Biochemistry*. The MacMillan Co., New York 1951, p. 429–430.

2. Oakley, C. L., Warrack, G. H., and Warren, M. E. The kappa and lambda antigens of *Clostridium welchii*. *J. Pathol. Bact.* **60**:495–503, 1948.

3. Lennette, E. H., Spaulding, E. H., and Truant, J. P. *Manual of Clinical Microbiology*, 2nd ed. American Society for Microbiology, Washington, D.C., 1974, p. 912.

4. Lautrop, H. A modified Kohn's test for the demonstration of bacterial gelatin liquefaction. *Acta Pathol. Microbiol. Scand.* **39**:357–369, 1956.

5. Gorini, L. Le rôle du calcium dans l'activité et la stabilité de quelques protéinases bactériennes. *Biochem. Biophys. Acta* **6**:237–255, 1950.

6. Levine, M. and Carpenter, D. C. Gelatin liquefaction by bacteria. *J. Bact.* **8**:297–306, 1923.

7. Frazier, W. C. A method for the detection of changes in gelatin due to bacteria. *J. Infect. Dis.* **39**:302–309, 1926.

8. Clarke, P. H. and Cowan, S. T. Biochemical methods for bacteriology. *J. Gen. Microbiol.* **6**:187–197, 1952.

9. Kohn, J. A preliminary report of a new gelatin liquefaction method. *J. Clin. Pathol.* **6**:249, 1953.

10. Pickford, G. E. and Dorris, F. Micro-methods for the detection of proteases and amylases. *Science* **80**:317–319, 1934.

11. LeMinor, L. and Piechaud, M. Note technique. Une méthode rapide de recherche de la proteolyse de la gélatine. *Ann. Inst. Pasteur.* **105**:792–794, 1963.

12. Hoyt, R. E. and Pickett, M. J. Use of "rapid substrate" tablets in the recognition of enteric bacteria. *Am. J. Clin. Pathol.* **27**:343–352, 1957.

13. Blazevic, D. J., Koepcke, M. H., and Matsen, J. M. Incidence and identification of *Pseudomonas fluorescens* and *Pseudomonas putida* in the clinical laboratory. *Appl. Microbiol.* **25**:107–110, 1973.

14. Porres, J. M. and Harris, D. Rapid gelatin liquefaction test. *Am. J. Clin. Pathol.* **62**:428–430, 1974.

CHAPTER 12

GLUCONATE OXIDATION

The ability to oxidize gluconate has been found to be most useful in the differentiation of species of *Pseudomonas*. The fluorescent pseudomonads can break down glucose by oxidative means according to the following general sequence (1–3):

Glucose → gluconic acid → 2-ketogluconic acid →
2-keto-6-phosphogluconic acid → → ⋯ 3-P-glyceraldehyde

This metabolic pathway can be easily detected because 2-ketogluconic acid is a reducing substance. In the performance of the test for gluconate oxidation, the organisms are grown in the presence of potassium gluconate which is not a reducing compound. If this is converted to 2-ketogluconate, one can detect the presence of the latter by a simple test for reducing substances. Usually this is done by using a "Clinitest" tablet (Ames Co., Elkhart, Ind.). Other nonspecific methods, such as Benedict's solution, can also be used.

The overall reaction of the test is described as follows:

$$
\begin{array}{ccccc}
\text{COO}^- & & \text{COO}^- & & \\
| & & | & & \\
\text{H}-\text{C}-\text{OH} & & \text{C}=\text{O} & & \\
| & & | & & \\
\text{HO}-\text{C}-\text{H} & \xrightarrow{\text{oxidation}} & \text{HO}-\text{C}-\text{H} & + \ 2\,\text{Cu}^{2+} & \xrightarrow[\text{alkali}]{\text{heat}} \ \text{Cu}_2\text{O} \downarrow \\
| & & | & & \\
\text{H}-\text{C}-\text{OH} & & \text{H}-\text{C}-\text{OH} & \text{(Clinitest,} & \text{Orange} \\
| & & | & \text{Benedict's,} & \text{precipitate} \\
\text{H}-\text{C}-\text{OH} & & \text{H}-\text{C}-\text{OH} & \text{etc.)} & \\
| & & | & & \\
\text{CH}_2\text{OH} & & \text{CH}_2\text{OH} & & \\
\end{array}
$$

Gluconate   2-Ketogluconate
(reducing substance)

The presence of an orange precipitate (cuprous oxide), after the Clinitest tablet or Benedict's solution is added to the inoculated and incubated culture, indicates a positive test for oxidation of gluconate. The cupric ions in Clinitest or Benedict's solution are reduced to cuprous ions by the free ketone group of 2-ketogluconate. With Benedict's solution the alkali necessary for the reduction to take place is part of the solution, and the heat is obtained by boiling. Clinitest tablets contain cupric ions, and sodium hydroxide is the source of alkalinity. Boiling is not necessary with these tablets because they also contain citric acid which combines with the sodium hydroxide in solution to generate heat.

The gluconate substrate for the test is available in the form of tablets from Key Scientific Products Co., Los Angeles, and Pease et al. (1) found these superior to the broth described by Haynes (2). Gaby and Free (3) also used the Key tablets and found them to be satisfactory and far simpler to use than is broth. These workers emphasized, however, that the test must be inoculated heavily or false-negatives may occur. In our laboratory, we have also found that an incubation period of 24 hours is superior to the minimum of 12 hours recommended by Key.

REFERENCES

1. Pease, M., Malcolm, J., Cherniak, R., and Dunlop, S. An approach to the problem of differentiating pseudomonads in the clinical laboratory. *Am. J. Med. Tech.* **34**:35–40, 1968.

2. DeLey, J. Comparative carbohydrate metabolism and localization of enzymes in *Pseudomonas* and related microorganisms. *J. Appl. Bact.* **23**:400–441, 1960.

3. Wood, W. A. and Schwerdt, R. F. Carbohydrate oxidation by Pseudomonas fluorescens. I. The mechanism of glucose and gluconate oxidation. *J. Biol. Chem.* **201**:501–511, 1953.

4. Haynes, W. C. *Pseudomonas aeruginosa*—its characterization and identification. *J. Gen. Microbiol.* **5**:939–950, 1951.

5. Gaby, W. L., and Free, E. Differential diagnosis of pseudomonas-like microorganisms in the clinical laboratory. *J. Bact.* **76**:442–444, 1958.

CHAPTER 13

HIPPURATE HYDROLYSIS

The detection of an organism's ability to hydrolyze sodium hippurate has been a useful characteristic in distinguishing between groups of streptococci and in the more complete speciation of anaerobic bacteria.

The fact that bacteria were able to hydrolyze hippurate was known as early as the late nineteenth century. It was not until 1922, however, when Ayers and Rupp (1) were studying the differences between human and bovine sources of hemolytic streptococci that practical use was made of the hippurate hydrolysis test. From preliminary studies these investigators knew that certain bacteria were able to split hippuric acid into benzoic acid and glycocoll (glycine); they therefore evaluated the action of bovine and human strains of beta hemolytic streptococci on an enriched medium containing hippuric acid. In the early part of their study they incubated the organisms in a peptone-enriched buffered medium with 1% sodium hippurate, pH 7.2, for 7 days at 37°C. They used a control tube without sodium hippurate for each test organism as well. After 7 days the cultures were acidified and distilled, and the volatile benzoic acid was titrated. A formol titration was also performed to determine the amino acid nitrogen produced in the reaction. They expressed the hippurate hydrolysis reaction as follows:

$$C_6H_5CONHCH_2COOH + H_2O \xrightarrow{\text{hippuricase}}$$

Hippuric acid

$$C_6H_5COOH + NH_2CH_2COOH$$

Benzoic acid   Glycocoll (glycine)

They found that bovine beta hemolytic streptoccocci were able to hydrolyze hippuric acid and that the proportions of benzoic acid and glycine formed correlated with the predicted quantitites of these end products. Beta hemolytic streptococci from human sources were unable to hydrolyze hippuric acid. These findings were extremely important, since serological methods for separating beta hemolytic streptococci were not yet known.

Leuthardt (2) in 1951 summarized the existing body of knowledge about the enzyme that is responsible for the hydrolysis of hippurate. This enzyme was appropriately named hippuricase, although in earlier tissue studies it had been called histozyme. Hippuricase is widely distributed in nature. It has been found in bacteria other than group B streptococci, in fungi, and in the organs of higher animals such as kidney and liver. Extracts containing hippuricase can hydrolyze N-acylated compounds as well as hippuric acids; however, only the benzol derivative of natural L-amino acids are split. In 1973 Ferrieri, Wannamaker, and Nelson (3) studied hippuricase activity in group B streptococci and found that the enzyme was largely intracellular. Cell-free filtrates of the organism showed no hydrolytic activity, but live or heat-killed cells were able to hydrolyze hippuric acid. When cells were mechanically disrupted excellent hippuricase activity could be demonstrated. These investigators also devised a microtiter method for assaying hippuricase activity which might have a clinical application.

Ayers and Rupp (1), recognizing that distillation and titration procedures are time consuming in clinical work, proposed two simple tests for detecting benzoate, one of the two hydrolysis products. To implement these procedures they modified the broth so that it contained peptone, pepsin, calcium chloride, one drop of 1% ferric chloride with 1% sodium hippurate, and a final pH of 7.1. This broth produced better growth of most beta hemolytic streptococci. Strains of streptococci were inoculated into this medium and incubated 7 days at 37°C. After incubation, a portion of the culture was removed, and 7% ferric chloride was added. The culture was shaken and then allowed to stand for 5–10 minutes. When hippurate had been split the ferric ion combined with benzoate to form an insoluble precipitate, ferric benzoate. Cultures of streptococci which were unable to split hippurate became clear or were slightly opalescent 5–10 minutes after the addition of ferric chloride. If a beef-infusion peptone medium with hippurate was required to secure good growth of the streptococci or if phosphates were present in the medium, the ferric chloride solution had to be modified to 12% made in 0.2–0.25% aqueous hydrochloric acid. In the latter medium, when hippurate had not been hydrolyzed, clearing took longer after the addition of the ferric chloride.

The second method that Ayers and Rupp (1) used to demonstrate benzoate formation was the acid test. For this test 50% sulfuric acid was added to the culture; the mixture was shaken and allowed to stand. A white precipitate of benzoic acid was formed when at least one half of the hippurate present in the medium was hydrolyzed.

Since the detection of hippurate hydrolysis had proved so useful in separating bovine beta hemolytic streptococci from human strains of these organisms, Hajna and Damon (4) explored the usefulness of this test for differentiating between specis of *Aerobacter* (*Enterobacter*). For their study they incorporated 0.3% sodium hippurate into the chemically defined medium of Koser (5). By using this medium they were able to determine the organism's ability to use hippurate as a source of carbon as well as its ability to hydrolyze it. They found that of the organisms described as *Coli–Citrobacter–Aerogenes–Clocacae* groups only *A. aerogenes* could utilize the carbon of hippurate for growth. In the process of utilizing hippurate, this organism hydrolyzed it. They recommended that the detection of growth of the organism in this chemically defined medium with hippurate correlated so well with the ferric chloride test for benzoate that the latter procedure was unnecessary. Although these findings are of interest, probably the greatest contribution made by Hajna and Damon (4) relates to a better explanation of the ferric chloride test for hippurate hydrolysis. They pointed out the importance of having an excess of ferric ion so that any residual hippurate, which would also precipitate with the addition of ferric chloride, would redissolve in an excess of the reagent while the ferric benzoate would remain as an insoluble precipitate. The requirement of a careful balance of ferric chloride reagent and sample to be tested was stressed by these investigators.

Cowan (5) devised a microtest to determine hippurate hydrolysis. A 1% sodium hippurate was used for a substrate combined with a phosphate buffer, bromthymol blue indicator, and test organism in a total volume of less than 0.2 ml. A control tube was set up with each test containing water instead of the hippurate aliquot of the mixture. All tubes were incubated for 24 hours at 37°C. Organisms were considered positive for hippurate hydrolysis if the pH of the hippurate substrate tube was 0.4 pH units higher than the control. Cowan (6) stated that the microtest was easier to read than the reference method he used for the study (4).

Since the detection of hippurate hydrolysis did not prove particularily helpful in differentiating between genera of the Enterobacteriaceae and since serological methods became available for distinguishing groups of beta hemolytic streptococci, less emphasis was placed on the hippurate hydrolysis methods and their improvement for almost 20 years. In 1973 Facklam and his associates (7) revived an interest in the test, pointing out that hippurate hydrolysis used in conjunction with bile esculin and 6.5% sodium chloride would provide a presumptive identification for group A, B, and D streptococci. These investigators recommended heart infusion broth with 1% sodium hippurate for the test; the medium was dispensed in screw capped tubes which could be tightened after sterilization to prevent evaporation. After inoculation with the organism, the culture was incubated at 35°C for 42–66 hours. At this point the culture was centrifuged and 0.8 ml of the supernate removed for the test. To this aliquot 0.2 ml of 12% ferric chloride in dilute hydrochloric acid was added (5.4 ml of concentrated hydrochloric acid was added to 94.6 ml of distilled water). A heavy cloudy precipitate which persisted after 10 minutes indicated that sodium hippurate had been hydrolyzed. Through the use of tightly capped tubes the fine balance between the quantity of spent medium and the amount of reagent could be more carefully controlled.

One of the disadvantages of considering the use of hippurate hydrolysis as a presumptive test was the 42–66 hours of incubation which it required. This problem prompted us to explore a rapid test for hippurate hydrolysis. For the most part, the product detected to determine hippurate hydrolysis previously was benzoate (1, 4, 7). The detection of a pH shift of the substrate to alkaline when hippurate was hydrolyzed used by Cowan (4) probably demonstrated alkalinity caused by the amino acid product, glycine. Rather than attempting to determine minor pH changes, we (8) developed a rapid method for hippurate hydrolysis using a heavy inoculum of beta hemolytic streptococci in 0.5 ml of 1% aqueous sodium hippurate substrate. The mixture was heated in a heating block at 37°C for 2 hours. The presence of the hydrolysis product, glycine, was then detected by adding a small quantity of ninhydrin to the mixture. The tubes were reincubated for 10 minutes; a deep purple color developed with ninhydrin if hippurate was hydrolyzed by the beta hemolytic streptococcus. This test correlated well with the hippurate hydrolysis test of Facklam et al. (5) but was far more rapid and easier to read.

The ninhydrin used in the previously described test is a strong oxidizing agent which evokes oxidative deamination of the alpha amino group. In the reaction, ammonia is liberated as well as carbon dioxide, the corresponding aldehyde, and a reduced form of ninhydrin. The ammonia then reacts with the residual ninhydrin and the reduced ninhydrin, hydrindantin, to form a purple substance (9). The reaction is shown below:

$$+ \ NH_2-CHR-CO_2H \ \rightarrow$$

Ninhydrin

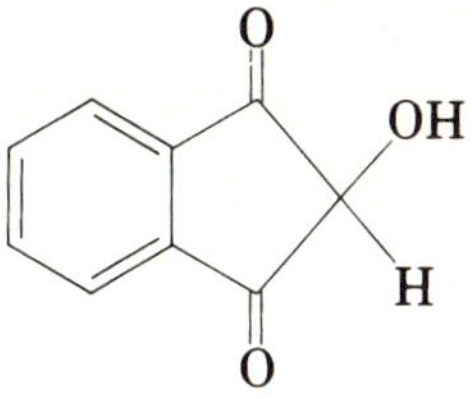

$$+ \text{RCHO} + CO_2 + NH_3$$

Hydrindantin

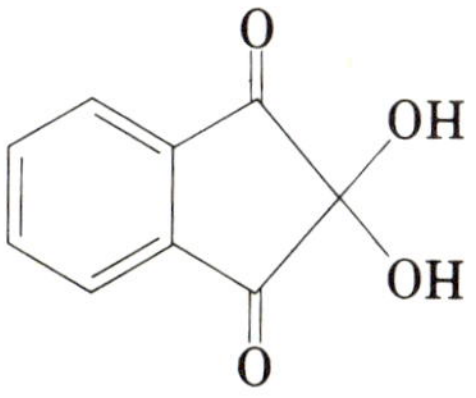

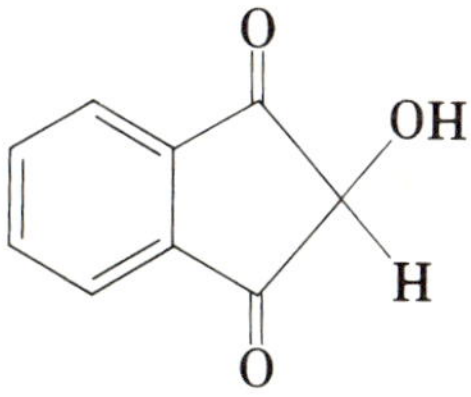

Ninhydrin   Hydrindantin $+ NH_3 \rightarrow$

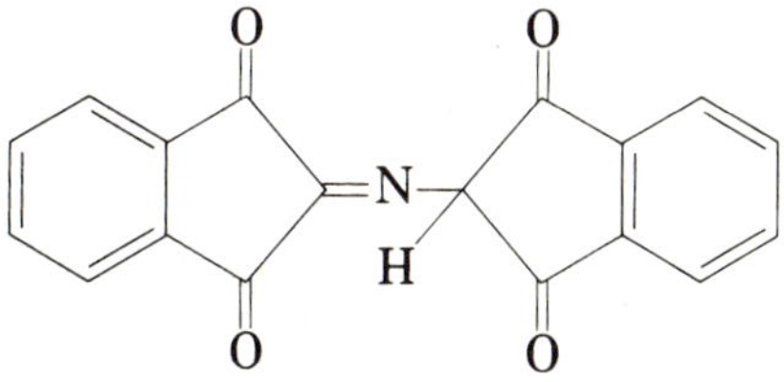

Purple

It was possible to use ninhydrin in the rapid hippurate hydrolysis test because the substrate contains only aqueous sodium hippurate which does not react with ninhydrin and because only one of the products, glycine, was an amino acid.

When hippurate hydrolysis tests are performed in the clinical laboratory using the ferric chloride method (1, 4, 7) it is important to bear in mind the fine balance of the medium substrate and the ferric chloride reagent required for this test. To control the constancy of the constituents of the medium with added hippurate the tubes should be tightly sealed after sterilization, preventing evaporation. It is helpful to include several uninoculated tubes of hippurate medium. These tubes should be incubated along with the test organisms to serve as one kind of negative control. If the ferric chloride reagent is added to one of these tubes and the ferric ion is in excess, the ferric hippurate formed at the moment of the addition of this reagent will disappear on standing for the appropriate time. If evidence of a negative reaction does not occur, the other tubes can be used to "titrate" the appropriate concentration of ferric chloride needed. Positive and negative control organisms should also be used for this test as well as for the rapid hippurate hydrolysis method employing ninhydrin for the detection reagent.

REFERENCES

1. Ayers, S. H. and Rupp, P. Differentiation of hemolytic streptococci from human and bovine sources by the hydrolysis of sodium hippurate. *J. Infect. Dis.* **30**:388–399, 1922.

2. Leuthardt, F., Hippuricase (Histozyme). In Summer, J. B. and Myrback, K., Eds. *The Enzymes: Chemistry and Mechanisms of Action,* Vol. 1, Academic Press, New York, 1951, pp. 951–955.

3. Ferrieri, P., Wannamaker, L. W., Nelson, J. A., Localization and characterization of hippuricase activity of group B streptococci. *Infect. Immun.* **7**:747–752, 1973.

4. Hajna, A. A. and Damon, S. R., Differentiation of *A. aerogenes* and *A. cloacae* on the basis of the hydrolysis of sodium hippurate. *Am. J. Epidemiol.* **19**:545–548, 1934.

5. Koser, S. A., Utilization of the salts of organic acids by the colon-aerogenes group. *J. Bact.* **8**:493–520, 1923.

6. Cowan, S. T., Biochemical test for bacterial characterization, 1. Hippurate Test. *J. Syst. Bact.* **5**:97–104, 1955.

7. Facklam, R. R. Padula, J. F., Thacker, L. G., Wortham, E. C., and Sconyer, B. J., Presumptive identification of group A, B and D streptococci. *Appl. Microbiol.* **27**:107–113, 1973.

8. Hwang, M. and Ederer, G. M., Rapid hippurate hydrolysis method for presumptive identification of group B streptococci. *J. Clin. Microbiol.* **1**:114–115, 1975.

9. Mahler, H. R. and Cordes, E. H., *Biological Chemistry*, 2nd ed. Harper and Row, New York, 1971, p. 54.

CHAPTER 14

HYDROGEN SULFIDE (H$_2$S) PRODUCTION

Bacteria may produce H$_2$S from organic sulfur compounds present in the peptones used in bacteriological media or from inorganic sulfur compounds added to culture media. The production of H$_2$S detects the organism's ability to reduce the sulfur to sulfide.

The production of H$_2$S from organic sulfur compounds is thought by some workers to be due to the action of an enzyme, cysteine desulfhydrase (previously called cysteinase), on the cysteine present in peptones (1,2). The overall reaction may be represented as:

$$HSCH_2CH(NH_2)COOH \xrightarrow[\text{desulfhydrase}]{\text{cysteine}} H_2S + NH_3 + CH_3COCOOH$$

Cysteine   Pyruvic acid

Cystine may also serve as a source for H$_2$S production, but it must first be reduced to cysteine.

Others have proposed that cysteine desulfhydrase is identical to cystathionase (3) and that the sulfur source is actually cystine rather than cysteine. Thus the production of H$_2$S from a defined substrate containing only cysteine may be due to preceding oxidation of the cysteine to cystine. However, these authors also state that other means of H$_2$S production by the intact cell may be operating.

As stated previously, H$_2$S can also be produced from an inorganic sulfur compound. Tilley (4) found that peptones varied widely in their content of reduceable sulfur. He recommended that the addition of thiosulfate as a sulfur source would make up for this variability and recommended its addition to media used for H$_2$S production. Media commonly used for detection of H$_2$S today, such as triple sugar iron agar (TSI) and Kligler's iron agar, contain sodium thiosulfate as the inorganic source of sulfur. Tarr (4) found that the production of H$_2$S from thiosulfate was due to a different mechanism than for the formation of H_2S from cysteine. It is thought that thiosulfate is reduced to sulfide in the presence of a bacterial enzyme (possibly a thiosulfate reductase) and a reducing agent which can act as a hydrogen donor (6,7). Substances such as cysteine and glutathione can act as the hydrogen donor. The general reaction is

$$S_2O_3^= + 2\,H \rightarrow SO_3^= + H_2S$$

Because different enzymes are responsible for the production of H_2S from either organic or inorganic sulfur compounds, results of tests for H_2S may vary depending on the type of medium or substrate used. For example, *E. coli* does not produce H_2S in TSI medium, but may produce H_2S in a medium containing large amounts of cysteine. It is obvious, therefore, that the results of tests for H_2S production can be compared only if the basal media are similar.

Another variable in tests for H_2S is the type of indicator used to detect the H_2S. The sulfide salts of many heavy metals such as lead, bismuth, and iron are black compounds, and salts of each of these metals have thus been used to detect H_2S production by bacteria, with the formation of a black color in the medium considered a positive test for H_2S. ZoBell and Felthan (8) compared the relative merits of each of these metals as detectors of H_2S. They found that when incorporated into media, iron salts were the most satisfactory and did not inhibit bacterial growth. Media such as TSI and Kligler's iron agar contain ferrous salts. The ferrous ions react with the H_2S to form the black precipitate, ferrous sulfide. Media containing ferric salts also show a black precipitate upon the formation of H_2S because after autoclaving of the medium and growth of the organism both ferrous and ferric ions are present because of changes in oxidation-reduction potential (8).

Although lead acetate was a very sensitive detector of H_2S, its presence in media inhibited the growth of some bacteria (8). However, when lead acetate-impregnated paper was inserted above the medium, this was the most sensitive method for detecting H_2S. In this test, the volatile H_2S produced by the bacteria reacts with the lead acetate on the paper to form the black precipitate, lead sulfide. The lead acetate paper method remains today the most sensitive means of determining H_2S production, when a suitable source of reduceable sulfur is present in the medium. The possibility does exist, however, that the extreme sensitivity shown by the lead acetate method may result from lead acetate's reacting with products of sulfur metabolism other than H_2S. Rodler et al. (9) demonstrated that with *E. coli* the black color obtained with lead acetate was due to the interaction of this indicator with mercaptanes produced from cystine. They concluded that for detection of H_2S only indicators such as iron salts, which give no reactions with mercaptanes, should be used. Their study shows a need for further examination of the use of lead acetate as an H_2S indicator.

It must be stressed that detection of H_2S production is not an absolute. Many organisms produce H_2S if a very sensitive method is used. For example, probably all members of the Enterobacteriaceae possess the capacity to produce H_2S although it may not be detected by all methods. It is essential in the differentiation of any organisms to use the method recommended for the particular group of organisms under study.

REFERENCES

1. Fromageot, C. Desulfhydrases. In Sumner, J. B., and Myrback, K., Eds. *The Enzymes. Chemistry and Mechanism of Action*, Vol. 1, Part 2. Academic Press, New York, 1951, p. 1237.

2. Smythe, C. V. Desulfhydrases and Dehydrases. In Colowick, S. P. and Kaplan, N. O., Eds. *Methods in Enzymology*. Vol. 2. Academic Press, New York, 1955, p. 315.

3. Cavallini, D., Mondovi, B., and De Marco, C. The identity of cysteine desulfhydrase with cystathionase and mechanism of cysteine-cystine desulfhydration. In Snell, E. E., Fasella, P. M., Braustein, A., and Fanelli, A. R., Eds. *Chemical and Biological Aspects of Pyridoxal Catalysis*. Pergamon Press, Oxford, 1963, p. 361.

4. Tilley, F. W. The relation between chemical composition of peptones and hydrogen sulfide production by bacteria. *J. Bact.* **8**:287–295, 1923.

5. Tarr, H. L. A. The enzymic formation of hydrogen sulfide by certain heterotrophic bacteria. II. *Biochem. J.* **28**:192–198, 1934.

6. Ishimoto, M., Kayama, J., and Nagai, Y. Biochemical studies on sulfate-reducing bacteria, IV. Reduction of thiosulfate by cell-free extract. *J. Biochem. (Tokyo)* **42**:41–53, 1955.

7. Kaji, A., and McElroy, W. D. Mechanism of hydrogen sulfide formation from thiosulfate. *J Bact.* **77**:630–637, 1959.

8. ZoBell, C. E., and Feltham, C. B. A comparison of lead, bismuth, and iron as detectors of hydrogen sulfide produced by bacteria. *J. Bact.* **28**:169–176, 1934.

9. Rodler, M., Vadon, V., and Pekar, K. Untersuchung des H₂S-Bildungsvermogens verschiedener Bakterien. *Zentralbl. Bakt.* **206**:117–122, 1968.

CHAPTER 15

INDOLE TEST

The ability of an organism to produce indole has long been used as a part of the IMViC reactions to separate *Escherichia* from *Klebsiella-Enterobacter* organisms, although it is now considered important in the identification of a wide variety of organisms isolated in the clinical laboratory.

Organisms produce indole from the amino acid tryptophane by the means of a tryptophanase (1). The source of the tryptophane can be the peptone in the culture medium used for growth, although it is recommended that the organisms be grown in a broth containing tryptone (2) which has an especially high tryptophane content. It is also recommended that the culture medium be devoid of glucose, since the presence of glucose appears to inhibit indole production (3, 4). The presence of oxygen influences the production of indole. Facultative organisms, such as *E. coli,* produce more indole when incubated aerobically (3, 4).

The indole produced by the organism can be detected with several different reagents, but the basic chemical reaction involved is the same. The Ehrlich-Boehme reagent (5) consists of two parts, *p*-dimethylaminobenzaldehyde-HCL in alcohol and potassium persulfate; each reagent is added, and a red color (rosindole dye) develops if indole is present. An alternative procedure is to add a solvent, such as xylene, ether, or chloroform, to extract and concentrate the indole; to this is then added the *p*-dimethylaminobenzaldehyde reagent, and a red color is formed in the solvent layer if indole is present. The latter procedure with xylene is the one usually used in the Ehrlich test.

Kovacs (6) modified the Ehrlich-Boehme reagent by using amyl alcohol in place of ethyl alcohol and found that extraction and concentration of indole occurred with the adding of the one reagent. Kovacs' reagent is probably the most widely used today.

In 1956 Gadebusch and Gabriel (7) reported that the substitution of isoamyl alcohol made the Kovacs' reagent more stable. Butyl alcohol is also satisfactory for the preparation of Kovacs' reagent (2).

The chemical reaction for detection of indole is based on the fact that when a pyrrole (indole = benzopyrrole) and a weakly acid alcoholic solution of p-dimethylaminobenzaldehyde are mixed (in the presence of heat), a red-violet color develops (8). The reaction will occur without heat if the reagent is made with concentrated HCl as are the currently used reagents.

The following reactions are involved:

$$\text{—CH}_2\text{CH(NH}_2\text{)COOH} + \text{H}_2\text{O} \xrightarrow{\text{tryptophanase}}$$

N
H

Tryptophane

$$+ \text{CH}_3\overset{\text{O}}{\overset{\|}{\text{C}}}\text{COOH} + \text{NH}_3$$

N
H

Indole   Pyruvic   Ammonia
acid

CHO

$$+ \quad \xrightarrow[\text{+H}^+]{\text{—H}_2\text{O}}$$

N
H   N(CH$_3$)$_2$

Indole p-Dimethylamino-
benzaldehyde

C

N   N
H   H

$$\text{N}^+(\text{CH}_3)_2$$

Rosindole dye

Although not shown in the reactions above, α-methylindole (indole-acetic acid) may also be produced from tryptophane, and this substance will react with the Ehrlich-Böhme or Kovacs' reagents to form the rosindole dye. Therefore, the previously described tests are not specific for indole itself. To make the test specific, advantage has been taken of the fact that indole is volatile whereas α-methylindole is not. Goré (9) modified the test for indole by not adding the reagent directly to the culture; he moistened the plug of the culture tube with the reagent, replaced the plug in the tube, and heated the culture. The volatile indole was demonstrated by the red color forming on the plug.

Holman and Gonzales (10) applied the oxalic acid reaction of Gnezda (11) to the detection of indole which is volatile at 37°C. In this test filter paper saturated with oxalic acid is placed in the top of the culture tube so that it does not touch the medium; as the volatile indole reaches the filter paper a pink color is formed. The chemical reaction between the indole and oxalic acid has not been described to date.

Because the Goré and Gnezda tests are specific only for indole, fewer positive reactions may be obtained using these procedures than with the Ehrlich-Böhme and Kovacs' tests.

Isenberg and Sundheim (12) investigated the specificity of the p-dimethylaminobenzaldehyde reagent for indole and found that it would react with 17 different indole compounds to form a red color after 5 minutes. The lack of specificity occurred when the reagent was added directly to the culture. If the culture fluid was first extracted with toluene and the reagent was then added to the toluene layer a red color was obtained only with indole and 5-methyl indole after 5 minutes.

Isenberg and Sundheim also studied the use of hydroxylamine hydrochloride to detect indole production by bacteria. This reagent reacted only with indole and 5-methyl indole even after 20 minutes. The chemistry of the reaction between indole and hydroxylamine is not known.

Many workers feel that Ehrlich's test is more sensitive than Kovacs'. When performing the test for indole on various groups of bacteria it is important that the procedure originally used in characterization of the organisms should be carried out. For example, with the Enterobacteriaceae Kovacs' reagent should be used, according to Ewing and Davis (13). On the other hand, when studying the nonfermenters (e.g., *Flavobacterium*) it is recommended that Ehrlich's reagent be used (14). Ehrlich's reagent is also recommended for detection of indole production in anaerobes (15).

Ordinarily organisms are inoculated into a broth high in tryptophane content, and the broth is incubated for at least 48 hours before performing the test for indole. However, a more rapid test may be used by taking a heavy inoculum from growth on a solid medium and inoculating a small amount (0.3–0.5 ml) of tryptone broth. The tryptophanase already present in the organisms will break down the tryptophane in the broth within 4 hours, and the indole may be detected in this short amount of time. In our laboratory this rapid test has given results equivalent to the standard longer procedures with the Enterobacteriaceae, *Flavobacterium*, and anaerobes. The commercially available reagent-impregnated strips also give results equal to those of the longer procedures with the Enterobacteriaceae, except for some strains of *Proteus rettgeri* (16), and anaerobes. A rapid spot test for indole production has also been used (17). In this test a piece of filter paper moistened with *p*-dimethylaminobenzaldehyde is placed in a Petri dish. A loopful of growth from a medium containing trytophane is placed on the filter paper; development of a red color within a few seconds indicates the presence of indole. With this method certain strains of *Proteus vulgaris, P. rettgeri, Providencia,* and *Aeromonas* will give a false-negative reaction.

REFERENCES

1. Gunsalus, J. C., Courtland G. G., and Stamer J. R. Tryptophane cleavage. In Colowick, S. P. and Kaplan N. O. *Methods in Enzymology,* Vol. 2. Academic Press, New York, 1955, p. 238.

2. *Manual of Microbiological Methods,* Society of American Bacteriologists, McGraw-Hill, New York, 1957, p. 155.

3. Suassana, I. and Suassana, I. R. The indole reactions of Enterobacteriaceae: II. Cultural conditions affecting indole production. *An. Microbiol.* **10**:123–145, 1962.

4. Fay, G. D. and Barry, A. L. Methods for detecting indole production by gram-negative nonsporeforming anaerobes. *Appl. Microbiol.* **27**:562–565, 1974.

5. Böhme, A. Die Anwendung der Ehrlichslen Indolreaktion für bakteriologische Zwecke. *Zentralbl. Bakl. Parasit. Abt. 1, Jena* **40**:129–133, 1906.

6. Kovacs, N. Eine vereinfachte Methode zum Nachwein der Indolbildung durch Bakterien. *Z. Immunitaets-forsch. Exp. Ther.* **55**:311–315, 1928.

7. Gadebusch, H. H. and Gabriel, S. Modified stable Kovacs' reagent for the detection of indol. *Am. J. Clin. Pathol.* **26**:1373–1375, 1956.

8. Feigl, F. *Spot Tests in Organic Analysis,* 7th ed. Elsevier Publishing Co., Amsterdam, 1966, pp. 199, 381.

9. Goré, S. N. A new technique of applying Ehrlich's reaction for detecting indole in bacterial cultures. *Ind. J. Med. Res.* **8**:505–507, 1921.

10. Holman, W. L. and Gonzales, F. L. A test for indol based on the oxalic acid reaction of Gnezda. *J. Bact.* **8**:577–583, 1923.

11. Gnezda, J. Sur des reactions nouvelles des bases indoliques et des corps albuminoides. *C. R. Acad. Sci.* **128:**1584–1587, 1899.

12. Isenberg, H. D. and Sundheim, L. H. Indole reactions in bacteria. *J. Bact.* **75:**682–690, 1958.

13. Ewing, W. H. and Davis, B. R. *Media and Tests for Differentiation of Enterobacteriaceae.* Center for Disease Control, Atlanta, Ga.1970.

14. King, E. O. *The Identification of Unusual Pathogenic Gram Negative Bacteria.* Center for Disease Control, Atlanta, Ga. 1967.

15. Holdeman, L. V. and Moore, W. E. C. *Anaerobe Laboratory Manual.* Virginia Polytechnic Institute and State University, Blacksburg, Va., 1972.

16. Blazevic, D. J., Ederer G. M., and Matsen, J. M. Evaluation of a new incubation-type indole strip test. *Appl. Microbiol.* **19:**547–548, 1970.

17. Vracko, R. and Sherris, J. C. Indole-spot test in bacteriology. *Am. J. Clin. Pathol.* **39:**429–432, 1963.

CHAPTER 16

LECITHINASE

The detection of lecithinase production by bacteria is of primary usefulness in identification of species within the *Clostridium* genus, although other organisms such as *Pseudomonas aeruginosa* or *Staphylococcus aureus* may also produce lecithinase (1).

Most of the work done in developing methods for lecithinase production and studying the mechanisms of the reaction have been carried out with *Clostridium perfringens* (*welchii*), and the discussion in this chapter is confined to the work with this organism, with mention of others where applicable.

In 1939 Nagler (2) studied the growth of *C. perfringens* in human serum and noticed that the serum became opalescent after 16 hours of incubation. The upper part of the serum was especially opaque. Nagler showed that the opalescence was caused by the lethal toxin of *C. perfringens* by demonstrating the inhibition of the reaction by specific antitoxin. This was the first report of this phenomenon in the literature, although Seiffert (3) also reported in 1939 on the same independently made observation. However, Seiffert stated that the reaction he observed was not inhibited by antitoxin.

MacFarlane and co-workers (4) followed up on the observation of Nagler and Seiffert, but also tested the effects of *C. perfringens* toxin on a crude lecithovitellin prepared from egg yolk. They found that the toxin caused opalescence in both serum and lecithovitellin. By the use of simultaneous tests for the hemolytic and lethal activities of the toxin and inhibition of these activities they concluded that (*a*) the opalescence was caused by α-toxin, (*b*) the opalescence was due to liberation of free fat from either the serum or lecithovitellin, and (*c*) the reaction with lecithovitellin was more rapid and more sensitive than that with serum.

Hayward (5) was the first to recognize the value of the opalescence in the identification of *C. perfringens*, and she referred to the test as the Nagler reaction, a name which has remained in use to this day. Of interest is that Hayward chose "Nagler reaction" even though Seiffert also described the reaction in 1939. She did this because Seiffert indicated the reaction was not inhibited by antitoxin; from this she concluded that the reaction Seiffert described was of a different nature. Hayward also developed an agar medium with human serum and noted that a zone of opacity would develop around colonies of *C. perfringens* growing on the plate (6). She tested other species of *Clostridium* and found that some of them also produced opaque zones and that the reaction with some of these organisms (e.g., *C. sordelli* and *C. bifermentans*) could be inhibited by *C. prefringens* antitoxin. Although egg yolk was a more sensitive substrate for the Nagler reaction, Hayward obtained better growth on media with serum and, therefore, recommended the use of serum rather than egg yolk. She also recommended the human serum plates for clinical specimens, since the Nagler reaction simplified the detection of *C. perfringens* in the presence of other organisms.

In 1947 McClung and Toabe (7) reported the development of an egg yolk agar which supported good growth of various clostridia and which gave more intense zones of opacity than serum media. Their medium is the most widely used one at the present time for detection of lecithinase activity. It is recommended that the eggs should be from chickens not receiving antibiotics to avoid possible inhibition of growth by antibiotic (8).

As mentioned previously, the Nagler reaction is caused by the action of the lethal, hemolytic toxin of clostridia upon serum or egg yolk. This toxin was subsequently shown to be a lecithinase by MacFarlane and Knight (9). They designated the enzyme from *C. welchii* as lecithinase C and found that it split lecithin into phosphorylcholine and a diglyceride. They also determined that the optimal pH for enzyme activity was 7.0–7.6 and that activity was stimulated by calcium ions.

In the studies described so far the investigators used a temperature of 37°C for the reaction. Theriault (10) found that the optimal temperature for activity of *C. perfringens* lecithinase was 45°C and recommended that tests for lecithinase potency be performed at the higher temperature. *C. perfringens* grows very well at 45°C, and thus a positive test for lecithinase at this temperature could be a specific means of identifying *C. perfringens*.

Although the enzyme responsible for the Nagler reaction is called a lecithinase C, it is also referred to as phospholipase C, phosphatidase, or phosphatidylcholine-phosphotidohydrolase and is in the general class of enzymes known as phosphodiester hydrolases (11). Lecithinase C hydrolyzes not only lecithin, but also other phospholipids, such as sphingomyelin, phosphatidylethanolamine, cephalin, and thromboplastin (8). It hydrolyzes lecithin (phosphatidyl choline) to produce phosphorylcholine and a diglyceride according to the following general reaction:

$$\begin{array}{l} H_2COOCR \\ | \\ R'COOCH \qquad O \\ | \qquad\qquad\quad \| \\ H_2C-O-P-OCH_2CH_2N^+(CH_3)_3 \\ \qquad\qquad | \\ \qquad\qquad O^- \end{array} \xrightarrow{\text{lecithinase C}}$$

Lecithin
(phosphatidylcholine)

$$\begin{array}{l} H_2COOCR \\ | \\ R'COOCH \\ | \\ H_2COH \end{array} \quad + \quad \begin{array}{l} O \\ \| \\ O=P-OCH_2CH_2N^+(CH_3)_3 \\ | \\ OH \end{array}$$

Diglyceride   Phosphorylcholine

Lecithinases (or phospholipases) A, B, and D have been described that also break down lecithin; however, they hydrolyze lecithin at different places on the molecule with the resulting production of substances other than phosphorylcholine and a diglyceride (12, 13).

The test for detection of lecithinase production by bacteria is usually performed by growing organisms on the egg yolk agar of McClung and Toabe or some modification thereof. After incubation, the presence of an opaque zone around the colonies indicates that lecithinase has been produced. The mechanism of the production of the opacity has been studied by Willis and Gowland (14). They presented evidence to show that the opalescence is due to a combination of three factors: (a) the fat (diglyceride) produced from the lecithin upon its breakdown by lecithinase, (b) free fats from the egg yolk emulsion after the breakdown of the lecithin (which is a stabilizer of the egg yolk emulsion), and (c) water-insoluble proteins from the egg yolk (vitellin and vitellenin) which precipitate with the free fats.

REFERENCES

1. Cowan, S. T. and Steel, K. J. *Manual for the Identification of Medical Bacteria*. Cambridge University Press, London, 1970, p. 31.

2. Nagler, F. P. O. Observations on a reaction between the lethal toxin of *Cl. welchii* (Type A) and human serum. *Br. J. Exp. Pathol.* **20:**473–485, 1939.

3. Seiffert, G. Eine Reaktion menschlicher Sera mit Perfringenstoxin. *Z. Immunitaetsforsch.* **96:**515–520, 1939.

4. MacFarlane, R. G., Oakley, C. L., and Anderson, C. G. Haemolysis and the production of opalescence in serum and lecitho-vitellin by the α-toxin of *Clostridium welchii. J. Pathol. Bact.* **52:**99–103, 1941.

5. Hayward, N. J. Rapid identification of *Cl. welchii* by the Nagler reaction. *Br. Med. J.* **1:**811–814, 1941.

6. Hayward, N. J. The rapid identification of *Cl. welchii* by Nagler tests in plate cultures. *J. Pathol. Bact.* **55:**285–293, 1943.

7. McClung, L. S. and Toabe, R. The egg yolk plate reaction for the presumptive diagnosis of *Clostridium sporogenes* and certain species of the gangrene and botulinum groups. *J. Bact.* **53:**139–147, 1947.

8. Smith, L. D. S. and Holdeman, L. V. *The Pathogenic Anaerobic Bacteria.* Chas. C. Thomas, Springfield, Ill., 1968, p. 29.

9. MacFarlane, M. G. and Knight, B. C. J. G. The biochemistry of bacterial toxins. I. The lecithinase activity of *Cl. welchii* toxins. *Biochem. J.* **35:**884–902, 1941.

10. Theriault, E. J. The lecithinase activity of *Clostridium perfringens* toxin. *U.S. Publ. Health Rep. Supp.* **188:**1–25, 1945.

11. Ispolotovskaya, M. V. Type A *Clostridium perfringens* toxin. In Kodis, S., Montie, T. C., and Ajl, S. L., Eds. *Microbial Toxins,* Vol. 2A. Academic Press, New York, 1971, pp. 109–158.

12. White, A., Handler, P., and Smith, E. L. Principles of Biochemistry, 5th ed. McGraw-Hill, New York, 1973, pp. 584–586.

13. Schwertner, H. A. and Friedman, H. S. Changes in lipid values and lipoprotein patterns of serum samples contaminated with bacteria. *Am. J. Clin. Pathol.* **59:**829–835, 1973.

14. Willis, A. T. and Gowland, G. Some observations on the mechanism of the Nagler reaction, *J. Pathol. Bact.* **83:**219–226, 1962.

CHAPTER 17

MALONATE UTILIZATION

The ability of certain genera within the family Enterobacteriaceae to utilize sodium malonate as the source of carbon has been useful in the differentiation of the members of this group.

In 1923 Koser (1), while studying bacterial nutritional requirements with special emphasis on the utilization of common organic acids by members of the colon-aerogenes organisms, observed that aerogenes-like organisms grew more luxuriantly than coli-like organisms in a chemically defined medium in which sodium malonate provided the only source of carbon. At that time this observation did not appear to be useful for the differentiation of organisms. Ten years later Leifson (2) described a medium with a somewhat different chemical composition in which sodium malonate was incorporated and an indicator, bromthymol blue, was included. Using this medium, he found that *Aerobacter* (*Enterobacter*) grew well and the indicator changed from green to blue. *Escherichia* grew poorly leaving the indicator unchanged. He also found a close correlation between the evidence for malonate utilization by organisms, indicator changing from green to blue, with the ability of these organisms to produce acetyl methylcarbinol. Since at one time it had been thought that acetyl methylcarbinol was formed from malonic acid, Leifson tested for this substance as an end product of malonate utilization. He was unable, however, to detect the presence of acetyl methylcarbinol in any of the positive substrates.

Several investigators (3–5) found that the addition of yeast extract to the substrate was necessary to increase the growth of all organisms; therefore, all organisms grew, but only those that could utilize malonate produced alkaline reactions. Edward and Ewing (6) recommended an additional modification in 1962 including a trace amount of dextrose as well as yeast extract. The modified malonate broth, then, has a carbohydrate source for carbon as well as the organic acid source, malonate. In the presence of the trace amount of glucose in the modified malonate broth, all organisms that can multiply in this medium and ferment dextrose produce acid products, shifting the pH to acid, <6.0, demonstrated by the bromthymol blue indicator changing to yellow. Those organisms that are able to utilize malonate, producing alkaline products in the substrate, cause a pH shift in the medium to alkaline, >7.6, shown by the indicator changing to a deep blue color.

As with acetate and citrate utilization, the exact nature of the alkaline reaction produced by organisms able to utilize the carbon of malonate is not completely clear. The reaction may occur because carbon dioxide is generated during utilization of citrate. Carbon dioxide combining with water and the sodium ion from sodium malonate may form sodium carbonate which would be sufficiently alkaline to change the indicator, bromthymol blue, from green to deep blue. As indicated in earlier chapters, the alkaline reaction may, on the other hand, be produced by the remaining excess sodium ions combining with the hydroxyl ions from water, forming sodium hydroxide. Clarification of the exact mechanism of this reaction, the nature of the enzyme responsible, and the need for permease would be helpful and might well shed light on the utilization of other organic acids by bacteria.

REFERENCES

1. Koser, S. A. Utilization of organic acids by the colon-aerogenes group. *J. Bact.* **8:**493–520, 1923.

2. Leifson, E. The fermentation of sodium malonate as a means of differentiating *Aerobacter* and *Escherichia*. *J. Bact.* **26:**329–330, 1933.

3. Shaw, C. Distinction between *Salmonella* and *Arizona* by Leifson's sodium malonate medium. *Int. Bull. Bact. Nomencl. Taxon.* **6:**1–4, 1956.

4. Davis, B. R., Ewing, W. R., and Reavis, R. W. The biochemical reactions given by members of the *Serratia* group. *Int. Bull. Bact. Nomencl. Taxon.* **7:**151–157, 1957.

5. Ewing, W. H., Davis, B. R., and Reavis, R. W. Phenylalanine and malonate media and their use in enteric bacteriology. *Publ. Health Lab.* **15:**153–167, 1957.

6. Edwards, P. R. and Ewing, W. H. *Publ. Health Serv. Bull.* No. 734, 19, 1962.

CHAPTER 18

METHYL RED (MR) TEST

The methyl red test was first described in 1915 by Clark and Lubs (1), who found it useful in differentiating between the coli-aerogenes group of enteric bacteria. It had previously been shown (2, 3) that these two groups of organisms could be divided on the basis of CO_2/H_2 ratios produced when the organisms were grown in the presence of dextrose. When grown *in vacuo*, one group produced a ratio of 1.06 ("low ratio" organisms) and the other a ratio of 1.90–3.00 ("high ratio" group). These workers believed that the ratios depended on basic differences in metabolism between the coli-aerogenes groups, and thus would be an accurate means of differentiation. However, determination of gas ratios was not a convenient method for identifying large numbers of cultures. Therefore, Clark and Lubs applied themselves to the development of a simple method that would correlate with the differences in gas ratios. They noted that when the coli organisms ("low ratio") were grown in a dextrose broth, acid was formed until a pH between 4 and 5 was reached. At this point the organisms ceased activity, and the pH then remained stable despite continued incubation. On the other hand, the aerogenes group ("high ratio") did not cease activity and continued to metabolize, with the medium then becoming more alkaline (pH between 6 and 7). It was suggested that the aerogenes organisms can attack the acids produced in their fermentation of dextrose and increase the pH, whereas the coli organisms cannot utilize their acid end products (4). Others have felt that the higher pH obtained with the aerogenes group is due to the acetoin formed in its fermentation of dextrose. Acetoin is neutral, and thus the pH is higher than in the cultures of the coli organisms where the major end products of fermentation are acids (5).

More recent evidence (6) has shown that under anaerobic conditions, with glucose as the sole carbon source for energy, *E. coli* produces relatively large amounts of organic acids (lactic, acetic, formic, and succinic) via the catabolic pathway known as *mixed acid fermentation*. Most *E. coli* strains do produce a hydrogenylase that splits formic acid into CO_2 and H_2; however, the CO_2/H_2 ratio is low with these organisms because much of the CO_2 is reincorporated into the formation of succinic acid. Aside from breaking down formic acid, *E. coli* does not utilize the organic end products of glucose metabolism under anerobic conditions.

By contrast, under identical conditions *Enterobacter aerogenes* produces small amounts of acidic products, but large amounts of neutral products in the form of butylene glycol and ethanol via the pathway known as *butylene glycol fermentation*. The CO_2/H_2 ratio is high both because succinic acid is not produced and because additional CO_2 is produced from the fusion of two molecules of pyruvate in the reactions leading to the formation of butylene glycol. The end products of butylene glycol fermentation cannot be utilized by *E. aerogenes* under anaerobic conditions.

The previous processes have been described as taking place under anaerobic conditions. Under aerobic conditions both *E. coli* and *E. aerogenes* have the potential to attack their respective fermentation products. However, *E. coli* produces about 130 moles of acid per 100 moles of glucose fermented, whereas *E. aerogens* produces only 20 moles of acid from the same amount of glucose. Therefore, *E. coli* would have to neutralize by oxidation five times the amount of acid as *E. aerogenes* in order to raise the pH above 6. This would be a formidable task, indeed, for these organisms.

In the development of the methyl red test Clark and Lubs evaluated several factors. They determined that a concentration of at least 0.5% dextrose was optimum to assure that the supply of dextrose would not be completely used up with the result that, under the aerobic conditions of the test, the medium might revert to alkaline. The best proportions of peptone and buffer were also determined, and the resulting medium is now marketed commercially as MR-VP medium. (It is also the recommended medium for the Voges-Proskauer test which detects the presence of the butylene glycol pathway.)

For proper performance of the test, Clark and Lubs found, the broth must be incubated aerobically at 30°C for 5 days. Their explanation of the long incubation period was the need to give the aerogenes group of organisms sufficient time to attack their fermentation end products and raise the pH.

To determine the pH of the medium after incubation methyl red is used because this indicator is red at pH 4–5 and yellow at pH 6–7. A red color after addition of the indicator indicates a positive methyl red test (mixed acid fermentation), and yellow indicates a negative methyl red test (butylene glycol fermentation).

There have been reports in the literature that the methyl red test does not always work and that the colors obtained are not always clear cut. However, it must be remembered that for accurate results MR-VP medium must be used, and the cultures must be incubated for 5 days at 30°C, as suggested by Clark and Lubs, who made no apology for urging the long incubation period.

Recently Barry et al. evaluated a more rapid (18 hours) methyl red test which they claimed to be comparable to the 5 day test (7). They used a heavy inoculum in 0.5 ml MR-VP broth. The small volume makes available more atmospheric oxygen, which enables the MR-negative organisms to continue breakdown of the acidic products to acetoin and butylene glycol; hence the pH rises more rapidly than in deep tubes of larger volume. They also found that the reduction of the broth volume eliminated the need to incubate the test at 30°C and that it worked very well at 37°C.

REFERENCES

1. Clark, W. M. and Lubs, H. A. The differentiation of bacteria of the colon-aerogenes family by the use of indicators. *J. Infect. Dis.* **17**:161–173, 1915.

2. Rogers, L. A. Clark, Wm. Mansfield, and Davis, Brooke J. The colon group of bacteria. *J. Infect. Dis.* **14**:411–471, 1914.

3. Rogers, L. A. Clark, Wm. Mansfield, and Evans, Alice C. The characteristics of bacteria of the colon type found in bovine feces, *J. Infect. Dis.* **15**:99–123, 1914.

4. Ruchhoft, C. C., Kallas, J. G., Chinn, B., and Coulter, E. W. Coliaerogenes differentiation in water analysis. II. The biochemical differential tests and their interpretation. *J. Bact.* **22**:125–181, 1931.

5. Burrows, W. *Textbook of Microbiology.* W. B. Saunders Co., Philadelphia, 1963, p. 160.

6. Doelle, H. W. *Bacterial Metabolism.* Academic Press, New York, 1969, Chapter 6.

7. Barry, A. L., Bernsohn, K. L., Adams, A. P., and Thrupp, L. D. Improved 18-hour methyl red test. *Appl. Microbiol.* **20**:866–870, 1970.

NITRATE REDUCTION

Bacteria may reduce nitrate by more than one process (1). A process called assimilation has nitrate reduced to ammonia via a series of poorly understood reactions that include nitrite as a known intermediate. The ammonia produced is then available to the cells for the synthesis of amino acids and other nitrogenous compounds.

In another process called dissimilation (or respiration) nitrate (or nitrite) is used by the organism primarily for energy metabolism rather than synthesis of essential nitrogen components. In this case nitrate (or nitrite) serves as the final electron acceptor in the absence or critically reduced presence of free oxygen. Thus this process allows those aerobes capable of nitrate respiration to grow anaerobically (e.g., *Pseudomonas*).

The initial reaction of nitrate reduction for either assimilation or dissimilation is mediated by nitrate reductase (also known as nitratase) which is closely associated with the cellular components of the electron transport system. The overall reactions involved in nitrate reduction may be described as

$$NO_3^- \xrightarrow[\text{reductase}]{\text{nitrate}} NO_2^- \longrightarrow NH_3$$

Certain bacteria convert nitrate to gaseous end products such as nitrogen or nitrous acid. This process is referred to as denitrification. Its relationship to nitrate reduction according to current knowledge (2) can be seen in the following scheme:

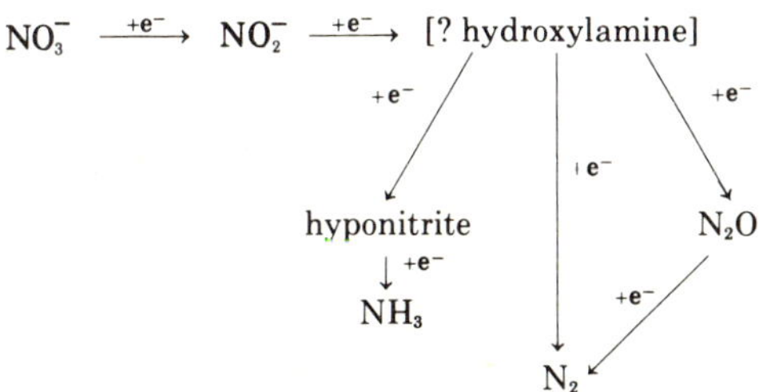

For determination of nitrate breakdown, organisms are usually grown in a suitable nutrient broth containing 0.1% KNO_3, and the presence of the resulting nitrite is determined by the addition first of sulfanilic acid and then of α-naphthylamine. The sulfanilic acid and nitrite react to form a diazonium salt (diazotized sulfanilic acid); the coupling of the diazonium salt with α-naphylamine then results in the production of a red water-soluble azo dye according to the following representation (3):

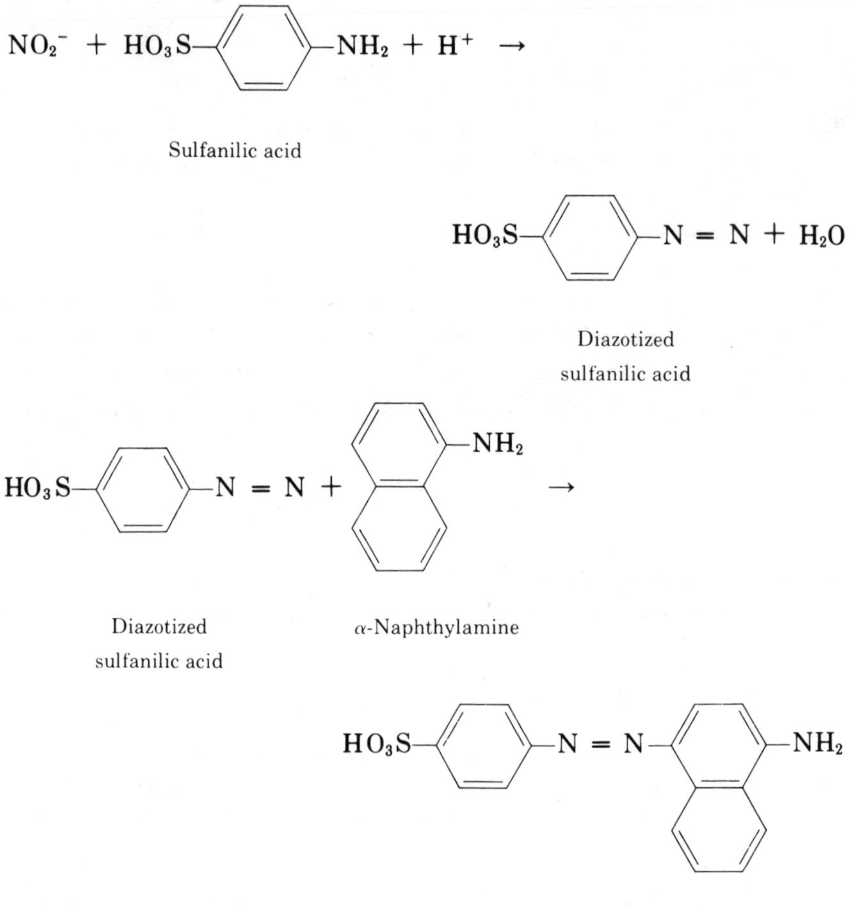

$$NO_2^- + HO_3S{-}\langle\bigcirc\rangle{-}NH_2 + H^+ \rightarrow$$

Sulfanilic acid

$$HO_3S{-}\langle\bigcirc\rangle{-}N = N + H_2O$$

Diazotized
sulfanilic acid

$$HO_3S{-}\langle\bigcirc\rangle{-}N = N + \text{(}\alpha\text{-naphthylamine)} \rightarrow$$

Diazotized
sulfanilic acid

α-Naphthylamine

$$HO_3S{-}\langle\bigcirc\rangle{-}N = N{-}\text{(naphthyl)}{-}NH_2$$

p-Sulfobenzeneazo-α-naphthylamine (red azo dye)

Thus the formation of a red color after the addition of the two reagents indicates that nitrite is present in the broth and is considered a positive test for nitrate reduction.

If a red color is not detected after the addition of the reagents, all available nitrate may have been reduced to nitrite, and then completely converted to ammonia, in which case no nitrite remains to react with the sulfanilic acid. It is also possible that dentrification has occurred with the complete conversion of the nitrate to nitrogen gas, again resulting in the absence of nitrite and thus no diazotization of the sulfanilic acid. The third possibility is that the nitrate was not reduced at all. To determine what has happened, a small pinch of zinc dust can be added to the tube. Metallic zinc reduces nitrate to nitrite. With the formation of nitrite upon the addition of zinc, the reactions outlined above can take place, resulting in the red color. A red color at this point indicates that nitrate was still present in the broth (a *negative* test for nitrate reduction). An absence of red color after the addition of zinc indicates that no nitrate was left in the broth and is thus a *positive* test for nitrate reduction. It is important not to add an excess of zinc when checking negative reactions. If too much of zinc is added, the large amount of hydrogen generated may reduce the nitrite (formed from unreduced nitrate) to ammonia; this could result in the lack of development of a red color or just a fleeting color reaction (4).

The production of gas from nitrate can also be determined by using an inverted Durham tube in the nitrate broth. If gas is produced, it will be trapped in the Durham tube, which also indicates a positive test for nitrate reduction. If gas bubbles are present, the nitrate reagents need not be added if the organism is a nonfermenter. If the organism is a fermenter, which could produce the gases hydrogen and CO_2 under the conditions of the test, the zinc must be added to prove the absence of nitrate. The production of gas may also be shown if a small amount of agar is added to the nitrate medium; bubbles or breaks in the agar indicate gas production.

The time at which the culture is tested for nitrate reduction is important; the organism in question must have grown well enough to have reduced the nitrate before the reagents are added. For most organisms an incubation time of 48 hours should be sufficient. A rapid test may also be performed by inoculating a large amount of growth into a small amount of nitrate broth; since the organism has already produced the enzymes necessary for reduction this test may be read after 2 hours of incubation at 37°C (4, 5).

If one wished to determine reduction of nitrite, the broth media inoculated may be prepared with the addition of KNO_2; a positive test for nitrite reduction is then indicated by a negative test for nitrite after incubation.

At times the red color may fade or change to a brownish color. This is due to the effect of excess nitrite upon the p-amino group of the azo dye. This color change can be reduced by the substitution of dimethyl-α-naphthylamine for the α-naphthylamine reagent (6).

Hugh (7) has suggested that 0.2% KNO_3,rather than 0.1%, be incorporated into the nitrate broth; it is then less likely that all the nitrite formed will be converted to ammonia, thus decreasing the number of occasions when zinc must be added to the test.

In the performance of the nitrate reduction test, one must be aware that contamination of glassware or reagents with nitrous oxide may give a false-positive test for reduction. This may be obviated by checking uninoculated tubes to be sure they give a negative reaction.

Although α-napthylamine has been the most widely used reagent for the nitrate reduction test, recent evidence has shown this agent to be carcinogenic. Therefore, the substitution of N,N-dimethyl-1-naphthylamine is recommended (8).

REFERENCES

1. Holding, A. J. and Collee, J. G. Routine biochemical tests. In Norris, J. R. and Ribbons, D. W., Eds. *Methods in Microbiology*, Vol. 6A, Academic Press, New York, 1971, p. 13.

2. Doelle, H. W. *Bacterial Metabolism*. Academic Press, New York, 1969, pp. 106–108.

3. Feigl, F. *Spot Tests in Organic Analysis*. Elsevier Publishing Co., Amsterdam, 1966, pp. 90–91.

4. Porres, J. M. and Porter, V. Rapid nitrate reduction test. *Am. J. Med. Tech.* **40:**257–259, 1974.

5. Blazevic, D. J. *Laboratory Procedures in Diagnostic Microbiology*. Telstar Productions, Inc., St. Paul, Minn., 1972, p. 109.

6. Wallace, G. I. and Neave, S. L. The nitrite test as applied to bacterial cultures. *J. Bact.* **14:**377–384, 1927.

7. Lennette, E. H. Spaulding, E. H., and Truant, J. P. *Manual of Clinical Microbiology*, 2nd ed. American Society for Microbiology, Washington, D.C., 1974, p. 263.

8. *Ibid.*, p. 934.

CHAPTER 20

ONPG TEST

Standard tests for lactose fermentation can demonstrate that lactose is hydrolyzed into its two monosaccharide components, galactose and glucose, but these tests are adequate only if the organism in question possesses both a lactose permease and a β-galactosidase. It is the enzyme β-galactosidase that catalyzes the breakdown of lactose; this intracellular enzyme has also been called "lactase" and is in the general class of enzymes called hydrolases. Some organisms produce this enzyme, but do not produce the permease that permits the entry of lactose into the cell. These organisms would thus yield negative results in a standard lactose fermentation test. The ONPG test is used to detect the presence of β-galactosidase in those organisms that lack the lactose permease.

Determination of β-galactosidase activity in lactose-negative organisms (organisms lacking permease) has been achieved by the use of O-nitrophenyl-β-D-galactoside (ONPG) as a substrate for the enzyme. ONPG is hydrolyzed by the same enzyme that hydrolyzes lactose (1), and was used in an extensive study of the β-galactosidase of *E. coli* by Lederberg (2). One advantage of using ONPG for demonstration of β-galactosidase is that ONPG is a colorless substance whereas one of the end products of its hydrolysis is O-nitrophenol, which is yellow. Thus the change from colorless to yellow indicates a positive test for β-galactosidase. The reaction involved is shown on page 84.

The optimal activity of the β-galactosidase when used as an extract occurs between pH 6 and 8 (3). The pH range when using intact cells may be wider. O-Nitrophenol is yellow at an alkaline pH. However, it is a weak acid, and the reaction solution must be well-buffered or have an alkaline pH because the undissociated acid is colorless. A pH of 7.5 is suitable to retain the O-nitrophenol in a yellow state and is also in the optimum pH range of the β-galactosidase.

A further advantage of using ONPG as substrate for β-galactosidase is that

ONPG
(colorless)

$+ H_2O \xrightarrow{\beta\text{-galactosidase}}$

Galactose O-Nitrophenol (yellow)

the rate of hydrolysis of ONPG is very high. In the presence of a large amount of enzyme the ONPG may be hydrolyzed in a few minutes, with the yellow color appearing within this time. Intact cells show good β-galactosidase activity; therefore by using a heavy inoculum of cells, which have already produced the enzyme, in a small volume of substrate the test will usually become positive in 1 hour.

There has been disagreement over whether β-galactosidase is a constitutive or inducible enzyme (2). However, it is now fairly well-accepted that β-galactosidase is an inducible enzyme, although it may be constitutive in some mutants of *E. coli* studied (4). It has been shown that glucose is an inhibitor to β-galactosidase activity and that cells grown on glucose show lower activity than those grown with lactose (2). The ONPG test as described by Ewing (5) recommends that organisms be cultured on a medium containing lactose (e.g., triple sugar iron agar).

REFERENCES

1.   Cohn, M. and Monod J. Purification et propriétés de la β-galactosidase (lactase) d'escherichia coli. *Biochim. Biaphys. Acta* 7:153–174, 1951.

2. Lederberg, J. The beta-*d*-galactosidase of *Escherichia coli,* strain K-12. *J. Bact.* **60**:381–392, 1950.

3. Wallenfels, K. and Malhotra, O. M. β-galactosidase In Boyer, P. D., Lardy, H., and Myrback, K., Eds. *The Enzymes,* Vol. 4. Academic Press, New York, 1960, p. 412.

4. Stanier, R. Y., M. Doudoroff, and E. A. Adelberg. *The Microbial World.* Prentice-Hall, Englewood Cliffs, N. J., 1970, pp. 282–284.

5. Ewing, W. H. and Davis, B. R. *Media and Tests for Differentiation of Enterobacteriaceae.* Center for Disease Control, Atlanta, Ga., 1970.

CHAPTER 21

OPTOCHIN INHIBITION

The susceptibility of pneumococci to low concentrations of ethyl hydrocuprein hydrochloride, optochin, is a useful test to distinguish pneumococci from alpha hemolytic streptococci.

Optochin was one of the early chemotherapeutic substances used in the treatment of pneumococcal pneumonia. It was introduced in 1911 by Morgenroth and Levy (1) who showed that this agent was effective in treating mice with pneumococcal infection. Shortly after Morgenroth and Levy published their findings, optochin was placed on the market and soon was being prescribed for treatment of acute lobar pneumonia in man. During the first few years of its use, little was known about the proper dosage, best route of administration, or possible adverse side effects of optochin. In 1917 Moore and Chesney (2) documented answers to these questions reporting on a study of 32 cases of acute lobar pneumonia that had been treated with optochin. They found that although optochin did produce bactericidal effects on pneumococci when administered at regular intervals orally, patients often suffered from serious toxic effects of the drug such as deafness and impairment of vision. In view of these findings, the use of optochin for the treatment of pneumococcal respiratory infections diminished. White (3) commented in his 1938 publication that an ideal drug for the treatment of pneumococcal infections had not yet been found.

While the efficacy of optochin for the treatment of pneumococcal infections was being evaluated a great number of *in vitro* studies were also being done. Shortly after optochin was placed on the market Moore (4) reported that very high dilutions of optochin inhibited the growth of pneumococci and that somewhat lower dilutions had bactericidal effects. Its effect on other bacteria was slight or absent, except on streptococci, and even then its inhibitory or bactericidal effect was considerably smaller on streptococci than on pneumococci. Moore at this early date stated that the *in vitro* activity of optochin aganist pneumococci helped to distinguish pneumococci from normal streptococcal flora of the mouth.

Forty years after Moore's report Bowers and Jeffries (5) introduced the use of optochin-impregnated disks into the field of clinical microbiology to facilitate distinguishing pneumococci in mixed or pure cultures. Each 8-mm sterile disk was impregnated with 0.02 ml of a 1:4000 dilution of optochin. With this concentration of optochin zones of inhibition for pneumococci ranged from 5 to 9 mm from the edge of the disk. Streptococci were not inhibited and grew right up to the edge of the disk or, in a few cases, produced small zones of 1 or 2 mm. Lund (6) evaluated a disk and tablet method for optochin sensitivity testing. The disk size was smaller in this study—6 mm; they were moistened in a 0.05% solution of optochin. The total zone of inhibition was measured; pneumococci produced zones of inhibition greater than 20 mm while streptococci produced no zone or zones of inhibition less than 15 mm. Since reports of the use of optochin for identifying pneumococci did not include information on the specific types of pneumococci that had been tested Bower, Thiele, Stearman, and Schaub (7) evaluated commercially prepared optochin disks (BBL) with 20 different serologic types of pneumococci. They found that optochin produced total zones of inhibition ranging from 15 to 30 mm when a heavy inoculum of pneumococci was used, whereas, with alpha streptococci a zone of inhibition, if it was produced, did not exceed 11 mm. There was no correlation between zones of inhibition and serologic type of pneumococcus.

Probably more investigative work would have been done on the mechanism of action of optochin had it been a more successful therapeutic agent. The structured formula is shown below:

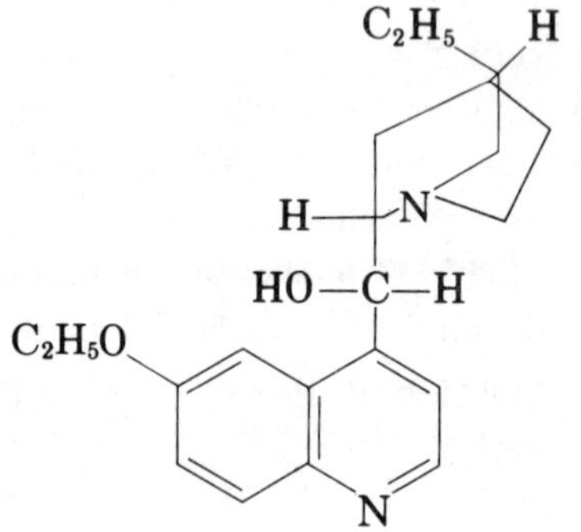

$$\text{optochin structural formula}$$

The structural formula of optochin bears a resemblance to quinine; however, the mode of action of quinine has not been completely described. It is possible, that optochin may interfere with the synthesis of folic acid as is the case with the sulfonamides and one group of antimalarial drugs (8). No studies have been made to determine if optochin acts as a competitive inhibitor of p-aminobenzoic acid.

When testing suspect pneumococci colonies for susceptibility to optochin, it is important to bear in mind the recent report of Ragsdale and Sanford (9). These investigators found that the zones of inhibition were considerably decreased when optochin susceptibility tests were incubated in a 5% CO_2 atmosphere rather than room-air atmosphere. Using the criterion of greater than 16 mm to indicate susceptibility, they found that the presence of pneumococci would not have been reported in 59% of 27 isolates. Although a small percentage of pneumococci require CO_2 for initial isolation (10) tests for susceptibility to optochin should be carried out in a room-air incubator.

REFERENCES

1. Morgenroth, J. and Levy, R. Chemotherapy of pneumococcal infection. *Berl. Klin. Wochenschr.* **48**:1560–1561, 1979–1983, 1911.

2. Moore, H. F. and Chesney, A. M. A study of ethylhydrocuprein (optochin) in the treatment of acute lobar pneumonia. *Arch. Int. Med.* **19**:611–682, 1917.

3. White, B. *The Biology of Pneumococcus*. The Commonwealth Fund, New York, 1938.

4. Moore, H. F. The action of ethylhydrocuprein (optochin) on type strains of pneumococci in vitro and in vivo, and on some other microorganisms in vitro. *J. Exp. Med.* **22**:269–285, 1915.

5. Bowers, E. F. and Jeffries, L. R. Optochin in the identification of *Str. pneumoniae*. *J. Clin. Pathol.* **8**:58–60, 1955.

6. Lund, E. Diagnosis of pneumococci by the optochin and bile tests. *Acta Pathol. Microbiol. Scand.* **47**:308–315, 1959.

7. Bowen, M. K., Thiele, L. C., Stearman, B. D., and Schaub, I. G. The optochin sensitivity test: a reliable method for identification of pneumococci. *J. Lab. Clin. Med.* **49**:641–642, 1957.

8. Goodman, L. S. and Gilman, A. The Pharmacological Basis of Therapeutics. Macmillan Co., London, 1970, p. 1098.

9. Ragsdale, A. R. and Sanford, J. P. Interfering effect of incubation in carbon dioxide on the identification of pneumococci by optochin discs. *Appl. Microbiol.* **22**:854–855, 1971.

10. Austrian, R. and Collins, P. Importance of carbon dioxide in the isolation of pneumococci. *J. Bacteriol.* **92**:1281–1284, 1966.

CHAPTER 22

OXIDASE TEST

The oxidase test is used chiefly in the identification of various genera of gram-negative bacteria.

In 1885 Ehrlich (1) first observed the oxidase reaction by injecting a mixture of α-naphthol and dimethyl-p-phenylenediamine into animals. He found that a blue substance, indophenol blue, was developed in the animal tissues at the site of injection. This reaction was referred to as the "nadi" reaction, from the first two letters of naphthol and diamine. In 1895 Rohrmann and Spitzer (2) studied this reaction in both plants and animals and found that it was brought about by an intracellular enzyme, which was called indophenol oxidase. The enzyme was also called "respiratory enzyme" by Warburg (3). Keilin in 1929 (4) demonstrated that indophenol oxidase and respiratory enzyme were identical. In 1938 Keilin and Hartree (5) discovered that the enzyme does not react directly with the p-phenylenediamine, but oxidizes cytochrome c, which in turn oxidizes the p-phenylenediamine. Hence the enzyme was renamed, "cytochrome oxidase." It is also referred to as "cytochrome c oxidase." Cytochrome c oxidase is now known to be a cytochrome complex, composed of cytochrome a and a_3, and is present in all aerobic respiratory bacteria including the Enterobacteriaceae. However, the actual substance reducing the oxidase reagent used with bacteria is oxidized cytochrome c, so that the oxidase test really determines the presence of cytochrome c and thus is positive only for bacteria containing cytochrome c as a respiratory enzyme. For this reason the test might more accurately be called the "cytochrome c test."

The cytochromes are heme-containing proteins and are oxidative enzymes in the respiratory chain of enzymes responsible for the reactions taking place in oxidative phosphorylation. In general, only bacteria that can utilize free oxygen as a terminal electron acceptor in their energy metabolism contain cytochromes; cytochromes are absent in most strict anaerobes of clinical importance. The iron in the heme of the cytochromes is responsible for the oxidized or reduced state of the cytochrome according to the following overall reaction:

$$\text{cytochrome–Fe}^{3+} + \text{e} \rightleftharpoons \text{cytochrome–Fe}^{2+}$$

ferric   ferrous

(oxidized)   (reduced)

The oxidation of cytochrome c, a necessary reaction in the development of a positive oxidase test, can be described as follows (6):

$$\textbf{2 reduced cytochrome c} + 2\text{ H}^+ + 1/2\text{ O}_2$$

$$\xrightarrow[\text{oxidase}]{\text{cytochrome}} \textbf{2 oxidized cytochrome c} + \text{H}_2\text{O}$$

In this reaction sequence the electrons from the reduced cytochrome c are taken up by oxidized cytochrome oxidase. Then the cytochrome oxidase passes these electrons to the oxygen molecule. Thus free oxygen is *necessary* for the *indirect* regeneration of oxidized-cytochrome c.

There are two main types of oxidase tests employed in clinical microbiology; they use different reagents, but the results and mechanism of action are comparable. The test first used with bacteria by Kovacs (7) has tetramethyl-*p*-phenylenediamine as a reagent which is easily oxidized by oxidized cytochrome c to form a purple substance according to the following reaction (8):

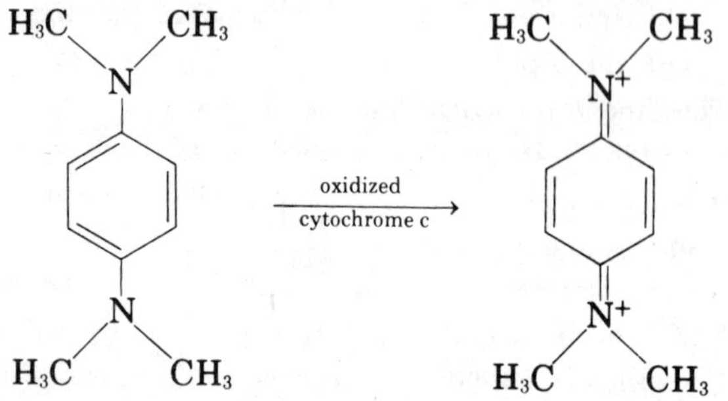

Tetramethyl-*p*-phenylenediamine   Wurster's blue

This procedure is usually performed by soaking a piece of filter paper with the reagent and then rubbing a portion of bacterial growth on the paper. Development of a purple color within 10 seconds is indicative of a positive oxidase test. It is important that a platinum wire or loop be used rather than nichrome because the nichrome itself can cause oxidation of the reagent.

The other common oxidase test is that of Gaby and Hadley (9) as modified by Ewing and Johnson (10). The reagents here are dimethyl-*p*-phenylene-diamine and α-naphthol (the original reagents used by Ehrlich). The reaction that takes place is as follows (8):

Dimethyl-*p*-phenylenediamine + α-Naphthol $\xrightarrow[\text{cytochrome c}]{\text{oxidized}}$

Indophenol blue $+ 4\,H^+ + 4\,e^-$

In this test the reagents are added to growth on a slant of nutrient agar, and the development of the blue color within 2 minutes indicates a positive test for cytochrome oxidase.

In performing either of these oxidase tests fresh reagents must be used, since both reagents become oxidized merely by the free oxygen in air.

Organisms containing a very small amount of cytochrome c may give a blue or purple color after longer periods than those stated for each test, and these organisms may be considered as weak-positive for oxidase. The times for development, however, must be adhered to, since a purple or blue color may develop later due to autooxidation of the reagents.

REFERENCES

1. Ehrlich, P. *Das Sauerstoff Bedurfnis des Organismus. Eine farbeanalytische Studie.* Hirschwald, Berlin, 1885.

2. Roöhrmann, F. and Spitzler, W. Über Oxydations-Wirkungen thierischer Gewebe. *Ber. Chem. Ges.* **28:**567–572, 1895.

3. Warburg, O. Über Eisen, den sauerstaffüberbertragender Bestandeil des Atmungsferments. *Biochem. Z.* **152:**479–494, 1924.

4. Keilin, D. Cytochrome and respiratory enzymes. *Proc. Roy. Soc. (London)* **B104:**206–251, 1929.

5. Keilin, D. and Hartree, E. F. Cytochrome oxidase. *Proc. Roy. Soc. (London)* **B125:**171–186, 1938.

6. Summer, J. B. and Somers, B. F. Chemistry and Methods of Enzymes, 2nd ed. Academic Press, New York, 1947, p. 225.

7. Kovacs, N. Identification of *Pseudomonas pyocyanea* by the oxidase reaction. *Nature* **178:**703, 1956.

8. Kersters, K. and DeLey, J. Enzymic tests with resting cells and cell-free extracts. In Norris, J. R. and Ribbons, D. W., Eds. *Methods in Microbiology,* Vol. 6A. Academic Press, New York, 1971, p. 35–36.

9. Gaby, W. L. and Hadley, C. Practical laboratory test for the identification of *Pseudomonas aeruginosa. J. Bact.* **74:**356–358, 1957.

10. Ewing, W. H. and Johnson, J. G. The differentiation of *Aeromonas* and C27 cultures from Enterobacteriaceae. *Int. Bull. Bact. Nomencl. Taxon.* **10:**223–230, 1960.

CHAPTER 23

PHOSPHATASE

Many bacteria produce phosphatases, but the test for phosphatase activity has been used chiefly to differentiate between pathogenic and nonpathogenic staphylococci, the former usually being phosphatase positive and the latter phosphatase negative (1–3). The production of phosphatase has also been proposed as a means of differentiating *Serratia* from *Enterobacter* (5).

Phosphatases are enzymes that hydrolyze esters of phosphoric acid. Those hydrolyzing monophosphoric esters are called phosphomonoesterases, and it is this type of phosphatase that is determined in the bacterial phosphatase test with staphylococci (6).

The general reaction of the monoesterases is as follows:

$$R-O-P(=O)(OH)-OH + H_2O \xrightarrow[\text{(phosphatase)}]{\text{monoesterase}} ROH + HO-P(=O)(OH)-OH$$

Monophosphoric ester　　　　　　　　　　　Alcohol　　Phosphoric acid

Several methods are available for determining phosphatase activity in microorganisms, all utilizing different substrates and different ways of detecting the end products of hydrolysis. The method originally described by Bray and King (7) utilizes phenolphthalein phosphate as substrate. Upon hydrolysis by the phosphatase, phenolphthalein is liberated. Phenolphthalein is a pH indicator that is colorless when acid and pink when alkaline. To detect the liberation of phenolphthalein alkali is added to the test, and the development of a pink color is indicative of a positive test for phosphatase. If the test is performed in a tube, sodium hydroxide is added to detect the phenolphthalein. In a test performed on a plate, after growth the colonies are exposed to ammonia vapors—colonies producing phosphatase turn pink because of the presence of the phenolphthalein.

Another test, described by Barnes and Morris (6), is a quantitative test for phosphatase production. An acid solution of p-nitrophenyl phosphate is here used as substrate. Upon hydrolysis p-nitrophenol is liberated, which becomes yellow when alkali is added.

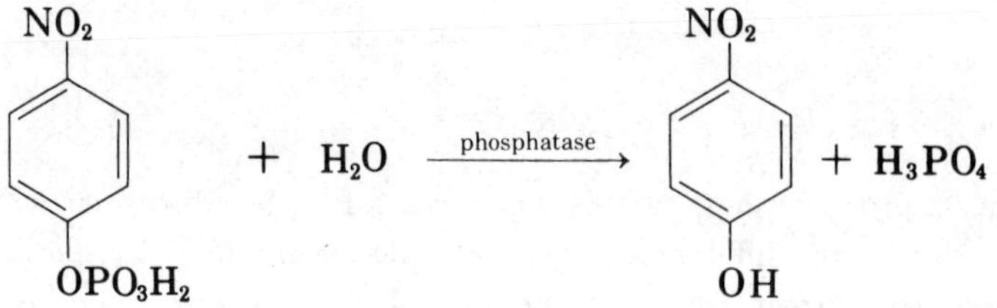

$$\text{p-Nitrophenylphosphate} + H_2O \xrightarrow{\text{phosphatase}} \text{p-Nitrophenol} + H_3PO_4$$

The enzyme determined by this method has a pH optimum of 5.6, hence is an acid phosphatase.

In a recently described test (4) the substrate for phosphatase activity is 5-bromo-4-chloro-3-indolyl phosphate at a pH of 5.2. Upon hydrolysis by acid phosphatase a blue-green indigo precipitate is formed:

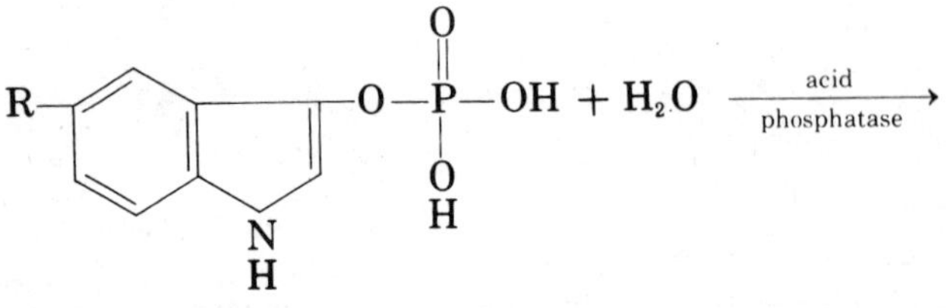

$$\text{Indolyl phosphate} + H_2O \xrightarrow[\text{phosphatase}]{\text{acid}} \text{Indolyl} + H_3PO_4$$

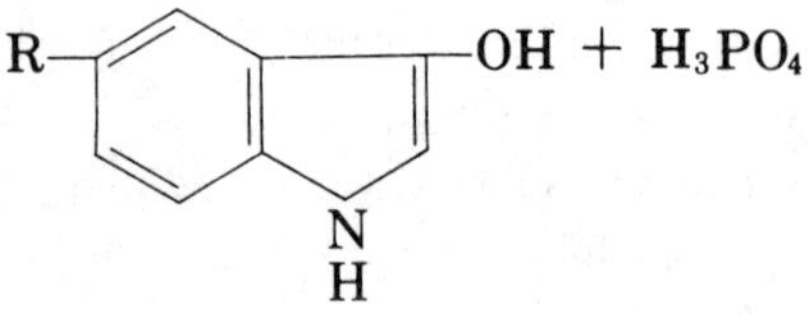

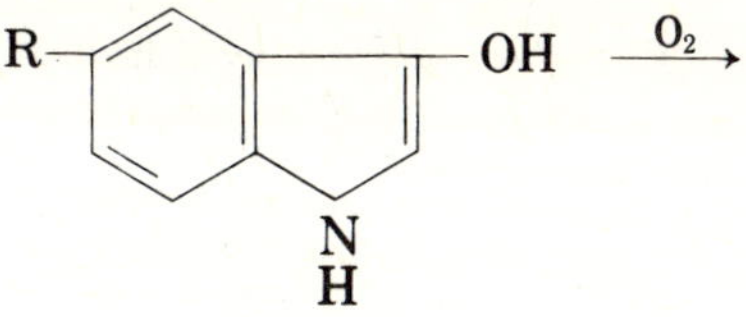

Indolyl

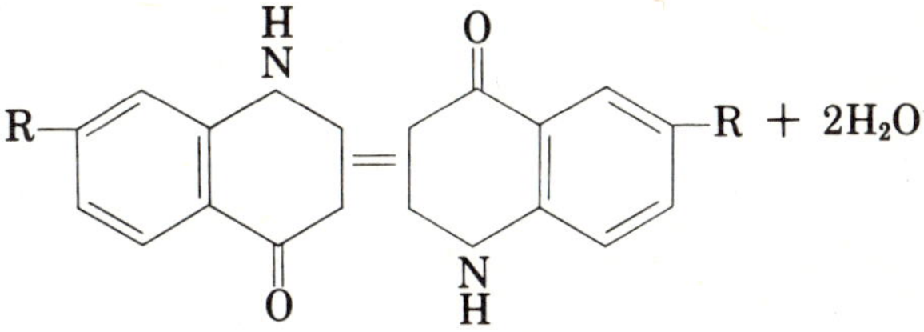

Blue-green indigo

This test is easy to read because of the dark color, but the substrate reagent is very expensive.

These two tests for acid phosphatase were compared by Kocka et al. (8) and were found to give equivalent results.

The tests described in this chapter detect nonspecific phosphatases as can be seen by the variety of substrates that can be used. There are many other bacterial enzymes that are phosphatases, but may require a specific substrate for demonstration of their activity.

REFERENCES

1. Barber, M. and Kuper, S. W. A. Identification of *Staphylococcus pyogenes* by the phosphatase reaction. *J. Pathol. Bact.* **63:**66–68, 1951.

2. Kedzin, W. The phosphatase activity of coagulase-positive *Staphylococcus aureus* strains isolated from patients and healthy persons. *J. Pathol. Bact.* **85:**528–529, 1963.

3. Pennoch, C. A. and Huddy, R. B. Phosphatase reaction of coagulase-negative staphylococci and micrococci. *J. Pathol. Bact.* **93:**685–688, 1967.

4. Von der Muehll, E., Ludwick, M., and Wolf, P. L. A new test for pathogenicity of staphylococci. *Lab. Med.* **3:**26–27, 1972.

5. Wolf, P. L., von der Muehll, E., and Ludwick, M. A new test to differentiate *Serratia* from *Enterobacter*. *Am. J. Clin. Pathol.* **57:**241 243, 1972.

6. Barnes, E. H. and Morris, J. F. A quantitative study of the phosphatase activity of *Micrococcus pyogenes. J. Bact.* **73**:100–104, 1957.

7. Bray, J. and King, E. J. Phenolphthalein phosphate as a reagent for demonstrating phosphatase in bacteria. *J. Pathol. Bact.* **54**:287, 1942.

8. Kocka, F. E., Magor, T., and Searcy, R. L. Evaluation of rapid tests for staphylococci characterization. *Am. J. Med. Tech.* **39**:269–271, 1973.

CHAPTER 24

STARCH HYDROLYSIS

The test for starch hydrolysis has been used with a wide variety of organisms, but today its greatest usefulness appears to be in the differentiation of various species of anaerobes.

Starch is composed of two polysaccharides, amylose and amylopectin, usually in a ratio of 1–4 to 1–5 (1,2). Starch granules heated in water (as in the production of a starch solution) swell, and the amylose fraction diffuses into the solution, whereas amylopectin remains insoluble. Both amylose and amylopectin are made up of D-glucose units. In amylose the glucose units are linked by 1,4-α-glucosidic bonds. In amylopectin the glucose units are linked by both 1,4- and 1,6-α-glucosidic bonds.

As early as 1918 Allen (3) reported on a simple method for detecting starch hydrolysis by microorganisms. The enzyme responsible for hydrolysis was called diastase. At the present time the enzyme responsible for starch hydrolysis is called amylase.

Amylases catalyze the hydrolysis of the 1,4-α-glucosidic bonds of polysaccharides such as starch (and glycogen). There are two types of amylases: alpha and beta. It is the α-amylase that is found in bacteria as well as in animal tissues (4). α-Amylase is also known as an endoamylase (5).

The hydrolysis of starch proceeds by first breaking the starch into dextrins and then into maltose and finally by converting the maltose to dextrose (6):

$$(C_6H_{10}O_5)n + nH_2O \xrightarrow{\alpha\text{-amylase}} (C_6H_{10}O_5)n \xrightarrow{\alpha\text{-amylase}}$$

$$\text{Starch} \qquad\qquad\qquad \text{Dextrin}$$

Maltose
(Glucose-4-α-glucoside)

Glucose
(α-D-glucose)

The actual mechanism of action of α-amylase is not completely known. The lack of knowledge is in part due to the complexity of the starch substrate, which is not a homogeneous compound, as stated previously. Also, in the series of hydrolytic reactions leading to glucose, the new products themselves become substrates for α-amylase. It is known, however, that chloride ion is necessary for activity of α-amylase; therefore, any test medium used to detect starch hydrolysis must contain chloride at a concentration of at least 0.01 M.

The optimal pH for α-amylase activity is in the range of 6.9–7.2. Although the optimal temperature for activity is 50–55°C, the reaction does appear to proceed satisfactorily at 37°C (4).

Detection of starch hydrolysis in bacteria usually makes use of an iodine reagent. The typical blue color obtained when iodine is added to starch is due to the amylose fraction. Amylose absorbs the iodine, and it appears that the iodine molecules fit into the helix of glucose units that make up amylose to form a blue inclusion compound (1). Amylopectin gives a red to violet color with starch.

The organisms are grown in a nutrient broth or agar containing starch. After incubation, an iodine solution (such as Gram's iodine) is added to the medium. A blue color indicates that starch is still present and that it has not been hydrolyzed. The absence of a blue color after addition of the iodine indicates an absence of starch, hence that the starch has been hydrolyzed. The color seen in a positive test may give a rough estimation of the extent to which the hydrolysis has progressed. If there are starch molecules left containing more than 30 glucose units a blue color will be obtained; the presence of 8–12 glucose units will give a red color, and glucose units of 6 or less will give no color (4). It is important that the reaction be read immediately after the addition of the iodine because the blue color formed with starch may fade.

Although most tests use an iodine reagent to determine the disappearance of starch, other methods have been employed (7). For example, 95% ethanol can be added to a plate culture after incubation; unhydrolyzed starch will be precipitated by the ethanol, giving a milky white appearance. If starch was hydrolyzed a clear area will appear around the organism. Another method detects the end product, glucose, by utilizing Benedict's reagent.

REFERENCES

1. Noller, C. R. *Chemistry of Organic Compounds,* 2nd ed. W. B. Saunders Co., Philadelphia, 1957, p. 396.

2. Geissman, T. A. *Principles of Organic Chemistry,* 2nd ed., W. H. Freeman and Co., San Francisco, 1962, p. 464.

3. Allen, P. W. A simple method for the classification of bacteria as to diastase production. *J. Bact.* **3**:15–17, 1918.

4. Henry, R. J. *Clinical Chemistry. Principles and Technics.* Harper & Row, New York, 1968, p. 469.

5. Fischer, E. H. and Stein, E. A. α-Amylases. In Boyer, P. D., Lordy, H., and Myrbach, K., Eds. *The Enzymes,* Vol. 4. Academic Press, New York, 1960.

6. Porter, J. R. *Bacterial Chemistry and Physiology.* John Wiley and Sons, New York, 1946, p. 497.

7. Cowan, S. T. and Steel, K. J. *Manual for the Identification of Medical Bacteria.* Cambridge Press, London, 1970.

CHAPTER 25

UREASE TEST

The test for the production of the enzyme urease by bacteria has been custo-marily used for the identification of organisms in the *Proteus* genus. However, many other medically important microorganisms also produce urease.

The reaction involved is based on the specificity of urease for urea as a substrate. Urea is hydrolyzed to form ammonia and carbon dioxide.

$$O=C\begin{array}{c} NH_2 \\ \\ NH_2 \end{array} + 2\,H_2O \xrightarrow{\text{urease}} 2\,NH_3 + CO_2 + H_2O$$

Urea   Ammonia
pH 6.8   pH 8.1

The ammonia formed causes the reaction mixture to become alkaline; therefore, an indication that the reaction has taken place (and that urease was present) can be achieved by the use of a pH indicator which shows a change in color from pH 6.8 to pH 8.1. The most commonly used indicator has been phenol red which has a pK of 7.9. At pH 6.8 phenol red is orangish; at pH 8.1 it is dark pink.

The sensitivity of the test for urease can be adjusted by the use of buffers. With a simple highly buffered medium Stuart *et al.* (1) found their test for urease to be specific for *Proteus* sp. Rustigian and Stuart (2) then modified this medium by decreasing the buffer content and found that a positive test specific for *Proteus* sp. could be obtained in 2–4 hours. *Proteus* characteristically produces larger amounts of urease than do most other organisms. Therefore, these tests are specific for *Proteus*, because the excessive amount of ammonia produced cannot be neutralized by the buffer content of these media.

Since it is useful to know about urease production in organisms other than *Proteus*, Christensen (3) devised another medium to detect small amounts of urease. The buffer content was reduced, and peptone and dextrose were added to permit more luxurious growth of the organisms being tested. These changes resulted in a medium that is extremely sensitive in detecting urease.

A number of other tests for urease production have been described (4–6). Their levels of sensitivity range between those of Rustigian and Stuart and of Christensen, depending primarily on their buffer content. The choice of test depends on the use one wishes to make of it.

Because all these tests for urease rely on the demonstration of alkalinity, they are not specific for urease. The pH might also be raised if the peptone or other proteins in the medium contained excessive amount of basic amino acid residues released upon protein hydrolysis. Certain organisms, such as *Pseudomonas aeruginosa*, may produce ammonia by breaking down peptones in the medium. In these cases a false-positive result for urease would be obtained. To determine if this is so, a control test on the same media not containing urea is necessary.

REFERENCES

1. Stuart, C. A., Van Stratum, E. and Rustigian, R. Further studies on urease production by *Proteus* and related organisms. *J. Bact.* **49**:437–444, 1945.

2. Rustigian, R. and Stuart, C. A. Decomposition of urea by *Proteus*. *Proc. Soc. Exp. Biol. Med.* **47**:108–112, 1941.

3. Christensen, W. B. Urea decomposition as a means of differentiating *Proteus* and paracolon cultures from each other and from *Salmonella* and *Shigella* types. *J. Bact.* **52**:461–466, 1946.

4. Ederer, G. M., Chu, J. H., and Blazevic, D. J. Rapid test for urease and phenylalanine deaminase production. *Appl. Microbiol.* **21**:545, 1971.

5. Matsen, J. M. Ten minute test for differentiating between *Klebsiella* and *Enterobacter* isolates. *Appl. Microbiol.* **19**:438–440, 1970.

6. Blazevic, D. J., Schreckenberger, P. C., and Matsen, J. M. Evaluation of the Pathotec "Rapid I-D System." *Appl. Microbiol.* **26**:886–889, 1973.

CHAPTER 26

VOGES-PROSKAUER (VP) TEST

This Voges-Proskauer test is used with the Enterobacteriaceae—organisms that have the ability to ferment glucose. In general, two pathways of glucose fermentation exist among the Enterobacteriaceae (1). Mixed acid fermentation results in large amounts of formic, acetic, lactic, and succinic acids and ethanol. Butylene glycol fermentation results in little or none of these acids, but a large amount of apolar butylene glycol and neutral ethanol. Acetylmethylcarbinol, or acetoin, is an intermediate in the production of butylene glycol. Since the VP test detects the presence of acetoin in the culture medium, it is the butylene glycol type of fermentation that is confirmed in the VP test.

In 1898 Voges and Proskauer (2) described a color reaction that they found useful in differentiating bacteria found in the intestine. A positive reaction occurred with those bacteria now known to be *Klebsiella* or *Enterobacter*, and a negative reaction took place with organisms now called *Escherichia coli*. These workers grew the organisms in a medium containing sugar and then added potash to the culture. After 24 hours or longer at room temperature a red color developed, indicating what they called a positive reaction.

In 1906 Harden (3) showed that the development of the red color was due to the production of acetylmethylcarbinol. In the presence of potash and air this substance is oxidized to diacetyl, which reacts with some constituent of the medium to form the red color. Diacetyl alone in the presence of potash does not yield a red color. Subsequently, a free NH_2 group on a guanidine moiety, such as that available from arginine present in peptone, was found to be the constituent necessary for the development of the color. O'Meara in 1931 (4) found that the color was intensified by the addition of creatine, which provides a greater amount of the guanidine groups than arginine on a gram-weight basis. With this modification, the red color developed within 15 minutes.

In 1936 Barritt (5) made the test more sensitive by adding α-naphthol. Without α-naphthol at least 10 ppm of diluted diacetyl was detectable; with α-

naphthol the sensitivity of the test was increased 50-fold. The exact role of the α-naphthol has not yet been defined.

It is important that the α-naphthol be added before the KOH. Barritt also found that meat infusion broth should not be used to grow the organisms because of the presence of acetoin, diacetyl, and related substances in the muscle extract which would give a false-positive test. The recommended medium for performance of the Voges-Proskauer test is commercially available as MR-VP medium (also called Clark-Lubs broth) and contains the optimal amounts of glucose, peptone, and buffer.

As the test is performed, α-naphthol and then KOH with creatine are added to the broth culture, and the tube is shaken. Acetoin is spontaneously oxidized in the presence of air and KOH to the intermediate, diacetyl, which in turn reacts with the guanidine group of creatine to form a red complex:

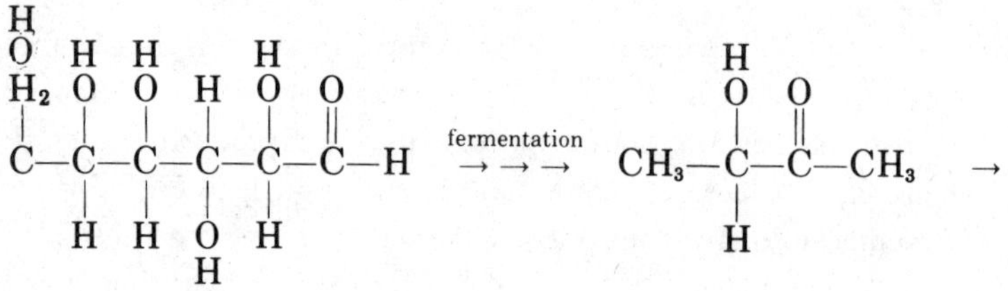

Glucose $\xrightarrow{\text{fermentation}}$ Acetylmethylcarbinol (acetoin) $\rightarrow$

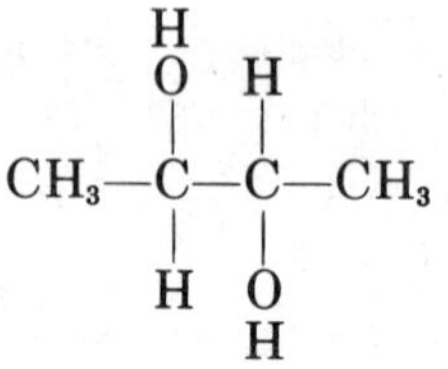

Butylene glycol (butanediol)

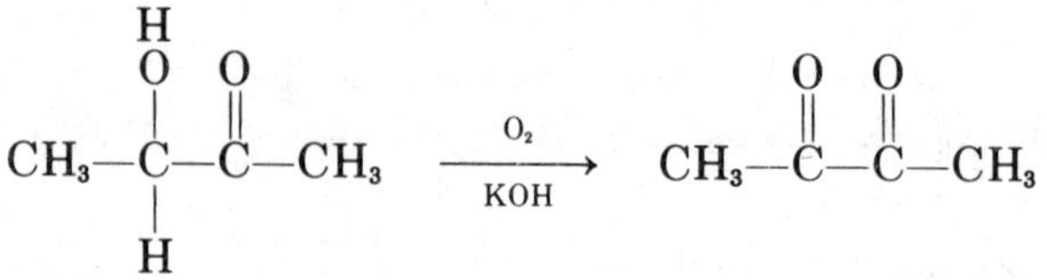

Acetylmethylcarbinol $\xrightarrow[\text{KOH}]{O_2}$ Diacetyl

$$CH_3-\overset{\overset{O}{\|}}{C}-\overset{\overset{O}{\|}}{C}-CH_3 + H_2N-\overset{\overset{NH}{|}}{C}-\underset{\underset{CH_3}{|}}{N}-CH_2COOH \rightarrow \text{red complex}$$

Diacetyl   Creatine

The chemistry of the color reaction is not known (6).

It has been recommended that the MR-VP medium be incubated for 48 hours before adding the reagents. However, rapid tests are available (7); these in general consist of inoculation of a heavy growth of organisms from a solid medium into a small amount (0.3–0.5 ml) of MR-VP broth. After incubation for 4 hours at 37°C sufficient acetoin has been produced to be detectable. If the organism to be tested has been grown on a medium containing glucose (e.g., MacConkey's agar, TSI), acetoin, if produced by the organism, may already be present, and the addition of the organism to the VP reagents will result in a red color immediately. Another rapid test utilizes reagent-impregnated strips; both of these rapid tests have been shown to give equivalent results to the standard 48-hour test (7, 8).

REFERENCES

1. Stanier, R. Y., Doudoroff, M., and Adeberg, E. A. *The Microbial World,* 3rd ed. Prentice-Hall, Englewood Cliffs, N.J., 1970, p. 183.

2. Voges, O. and Proskauer, B. Beitrag zur Ernährungsphysiologie und zur Differential-Diagnose der Bakterien der hämorrhagischen Septicamie. *Z. Hyg.* **28:**20–32, 1898.

3. Harden, A. On Voges and Proskauer's reaction for certain bacteria. *Proc. Roy. Soc. (London)* **77:**424–425, 1906.

4. O'Meara, R. A. Q. A simple delicate and rapid method of detecting the formation of acetylmethylcarbinol by bacteria fermenting carbohydrate. *J. Pathol. Bact.* **34:**401–406, 1931.

5. Barritt, M. M. The intensification of the Voges-Proskauer reaction by the addition of α-napthol. *J. Pathol. Bact.* **42:**441–453, 1936.

6. Feigel, F. and Anges, V. *Spot Tests in Organic Analysis,* 7th ed. Elsevier, New York, 1966, p. 445.

7. Barry, A. L., and Feeney, K. L. Two quick methods for Voges-Proskauer test. *Appl. Microbiol.* **15:**1138–1141, 1967.

8. Blazevic, D. B., Schreckenberger, P. C., and Matsen, J. M. Evaluation of the PathoTec "Rapid I-D System." *Appl. Microbiol.* **26:**886–889, 1973.

CHAPTER 27

TEST PROCEDURES

ACETATE UTILIZATION

Prepare acetate differential agar according to manufacturer's directions.

1. Prepare a saline suspension of the organism.
2. Inoculate the acetate medium slant with the saline suspension using a straight wire.
3. Incubate for 24–48 hours at 35°C (maximum time 7 days).

Positive: growth, blue color.
Negative: no growth, green color.

REFERENCE

Trabulsi, L. R. and Ewing, W. H. Sodium acetate medium for the differentiation of *Shigella* and *Escherichia* cultures. *Publ. Health Lab.* **20**:137–140, 1962.

BILE SOLUBILITY

Rapid Plate Method

Prepare either 2% sodium lauryl sulfate or 10% sodium desoxycholate. White Dreft, if available, can be used as an alternate in a 1% solution.

1. Place a drop of the reagent on suspected pneumococcal colonies. Allow the reagent to dry on the plate taking care not to disturb the drop.
2. After the drop of reagent has dried, look for disappearance of the colonies. Do not confuse colonies that have floated away with lysis.

Positive: lysis of colony.
Negative: no lysis.

REFERENCE

Hawn, C. V. Z. and Beebe, E. Rapid method for demonstrating bile solubility of *Diplococcus pneumoniae*. *J. Bact.* **60**:549, 1965.

Tube Method (Dreft)

Prepare veal infusion broth according to manufacturer's directions. Dispense in 5 ml aliquots. When sterile, add 0.1 ml horse serum to each tube.

1. Inoculate the broth-serum tube and incubate at 35°C until the tube is visibly turbid.
2. Add several drops of 4% Dreft (sodium lauryl sulfate). Observe to determine clearing of turbidity.

    Positive: rapid clearing of turbidity.
    Negative: no clearing.

REFERENCE

Washington, J. A., II. *Laboratory Procedures in Clinical Microbiology.* Little, Brown and Co., Boston, 1974, pp. 461–462.

CATALASE

Tube Test

1. Inoculate an agar slant heavily with the organism and incubate at an optimal temperature for 18–24 hours.
2. Introduce 1 ml of 3% hydrogen peroxide, allowing it to flow over the slant.

    Positive: gas bubbles produced.
    Negative: no gas produced.

Plate Test

1. Place a drop of 3 or 30% hydrogen peroxide on a slide.
2. Pick a single colony or growth from a slant and emulsify the growth in the drop of hydrogen peroxide. Colonies may be picked from blood containing medium despite the fact that blood contains the enzyme, catalase, if extreme caution is used.

    Positive: gas bubbles produced.
    Negative: no gas produced.

REFERENCE

Paik, G., and Suggs, M. T. Reagents, stains and miscellaneous test procedures. In Lennette, E. H., Spaulding, E. H., and Truant, J. P., (Eds.) *Manual of Clinical Microbiology* American Society for Miciobiology, Bethesda, Md., 1974, p. 931.

CITRATE UTILIZATION

Prepare Simmons' citrate according to manufacturer's directions.

1. Inoculate the surface of the slant lightly using a saline suspension of the organism and a straight wire.
2. Incubate for 24–28 hours (maximum 7 days) at 35°C.

    Positive: growth, alkaline reaction.
    Negative: no growth, no change of indicator.

REFERENCE

Simmons, J. S. A culture medium for differentiating organisms of the typhoid-colon-aerogenes groups. *J. Infect. Dis.* **39:**209–214, 1926.

COAGULASE TEST

Tube Test (Free Coagulase)

Rehydrate rabbit plasma according to manufacturer's directions.

1. Emulsify the organism in 0.5 ml of diluted plasma. Use positive and negative controls.
2. Incubate at 35°C in a water bath and observe at 4 hours to detect clot formation. If no clot is formed, reincubate and examine again at 18–24 hours.

    Positive: clot formation.
    Negative: no clot formed.

REFERENCE

Washington, J. A., II, Ed. *Laboratory Procedures in Clinical Microbiology.* Little, Brown and Co., Boston, 1974, p. 461.

Slide Test (Clumping Factor or Bound Coagulase)

Rehydrate rabbit plasma according to manufacturer's directions. Do not dilute. Dispense into a sterile bottle with a dropper pipette. Human plasma may also be used for the slide coagulase test.

1. Make a very heavy (milky) homogeneous suspension of staphylococci in a drop of water on a slide.
2. Add one drop of undiluted plasma to the suspension and mix thoroughly for 5 seconds.
3. Prepare a positive and negative control in the same manner.

   Positive: macroscopic clumping within 5–15 seconds.
   Negative: suspension remains smooth.

REFERENCE

Bailey, W. R. and Scott, E. G. *Diagnostic Microbiology,* 4th ed., C. V. Mosby Co., St. Louis, 1974, p. 112.

DECARBOXYLASE (LYSINE AND ORNITHINE) AND ARGININE DIHYDROLASE

Moeller's Method

Prepare Moeller Decarboxylase Broth Base according to manufacturer's directions, adding 1% L-lysine dihydrochloride, 1% L-ornithine dihydrochloride, or 1% L-arginine monohydrochloride. Also prepare Moeller's Decarboxylase Broth Base without the addition of amino acid (control).

1. Inoculate one tube of amino acid medium and one tube of control medium.
2. Add a layer of sterile mineral oil (4–5 mm) to each tube.
3. Incubate at 35°C and examine daily until test becomes positive or 4 days have passed.

   Positive: violet or red-violet color (with yellow in control).
   Negative: yellow color in both test and control.

REFERENCE

Moeller, V. Simplified tests for some amino acid decarboxylases and for the arginine dihydrolase system. *Acta Pathol. Microbiol. Scand.* **36:**158–172, 1955.

Falkow's Method

Prepare Decarboxylase Medium Base according to manufacturer's directions, adding 0.5% L-lysine dihydrochloride, 0.5% L-ornithine dihydrochloride, or 0.5% L-arginine monohydrochloride. Also prepare Decarboxylase Medium Base without the addition of amino acid. Dispense media in screw-capped tubes.

1. Inoculate one tube of amino acid medium and one tube of control medium.
2. Incubate at 35°C and examine daily until test becomes positive or 4 days have passed.

   Positive: purple color (control—yellow).
   Negative: yellow color in both test and control.

REFERENCE

Falkow, S. Activity of lysine decarboxylase as an aid in the identification of *Salmonellae* and *Shigellae. Am. J. Clin. Pathol.* **29**:598–600, 1958.

Lysine Iron Agar (LIA) Method

Prepare LIA according to manufacturer's directions. Slant so there is at least a 1-in. butt.

1. Inoculate LIA by streaking the slant and stabbing the butt twice.
2. Incubate at 35°C overnight.

   Lysine decarboxylase positive: purple butt.
   Lysine decarboxylase negative: yellow butt.

Motility-Indole-Ornithine (MIO) Method

Prepare MIO medium according to manufacturer's directions. Tube in 5-ml amounts in 13 × 100 mm tubes.

1. Inoculate MIO medium by stabbing to the *bottom* of the tube with a straight wire.
2. Incubate overnight at 35°C.

   Ornithine decarboxylase positive: purple or greyish color throughout tube.
   Ornithine decarboxylase negative: yellow color with narrow rim of purple at top of medium.

REFERENCE

Ederer, G. M., and Clark, M. Motility indole ornithine medium. *Appl. Microbiol.* **20:**849–850, 1970.

Thin-Layer Chromatography Method (Arginine Dihydrolase)

Prepare 0.2% L-arginine monohydrochloride in distilled water. Dispense in 0.25-ml amounts into 13 × 100 mm tubes; may be stored at −20°C until used.

1. Inoculate 0.25 ml of arginine substrate with a heavy loopful of growth from TSI or other solid medium.
2. Incubate in a heating block or water bath at 37°C for 2 hours. Also incubate an uninoculated tube of arginine substrate.
3. While the tubes are incubating mix 28 ml of Solution I and 12 ml of Solution II in a Desaga developing tank. Place a sheet of blotting paper on the two long sides of the tank to saturate the atmosphere.
4. Mark a sheet of Cel 300 TLC (MN Polygram Cel 300 cellulose sheets, Brinkman) lightly with a pencil line 1.5 cm from the bottom of the sheet. Draw a parallel line 10 cm above the first line. Do not touch the surface of the sheet with fingers. Place pencil dots at 1.0-cm intervals along the first line.
5. Remove the tubes from the heating block without agitation. Aspirate 2 μl from tube, using 2 μl disposable capillary pipettes. Deliver half the measured amount on one of the marks on the lower line on the sheet, dry with a stream of air; apply the remaining fluid to the same spot and dry with a stream of air.
6. Apply 2 μl of the uninoculated substrate to the sheet in the same manner. Also apply 2 μl of 0.2% L-ornithine dehydrochloride and 0.2% L-citrulline in two separate spots on the sheet.
7. Place the sheet in the Desaga tank, taking care that the sheet does not touch the blotters. Tightly cover the tank.
8. When the solvent has reached the 10-cm mark (about 45 minutes) remove the sheet from the tank. Dry in the air or on a slide dryer at 45–50°C for about 10 minutes.
9. Spray sheet to saturation with ninhydrin solution (7% ninhydrin in Solution I). Spray in a hood.
10. Dry on a slide drier at 45–50°C until purple spots appear (about 15–20 minutes).
11. Compare spots produced by organism to the standards.

Positive for arginine dihydrolase: ornithine and citrulline spots, or only ornithine.

Solution I: Mix 100 ml of *N*-butanol and 100 ml of acetone. Store at 4°C.
Solution II: Mix 50 ml of glacial acetic with 100 ml of distilled water.

REFERENCE

Williams, G. A., Blazevic, D. J., and Ederer, G. M. Detection of arginine dihydrolase in nonfermentative bacteria by use of thin layer chromatography. *Appl. Microbiol.* **22**:1135–1137, 1971.

DEOXYRIBONUCLEASE (DNase TEST)

Method 1: Gram-Negative Organisms Only

Prepare DNase Test Agar according to manufacturer's directions. Add 0.5 ml of 2% aqueous toluidine blue for each 200 ml of medium. Dispense in 2-ml aliquots into 13 × 100 mm tubes. Autoclave at 121°C for 15 minutes; slant after autoclaving. The medium may also be poured into plates after autoclaving if preferred.

1. Inoculate medium heavily in one spot; incubate overnight or longer at 35°C.
2. Read results as follows:

   Positive: Rose-pink zone around growth.
   Negative: No change in color or medium.

REFERENCE

Blazevic, D. J. Laboratory procedure in diagnostic microbiology. Telstar Productions, Inc., St. Paul, Minn., 1974, p. 133.

Method 2: Gram-Positive and Gram-Negative Organisms

Prepare DNase Test Agar according to manufacturer's directions. Pour into plates after autoclaving.

1. Inoculate medium heavily in one spot; incubate overnight or longer at 35°C.
2. Flood plate with IN HCl.

3. Read results as follows:

Positive: Clear zone around growth.
Negative: No clearing around growth.

REFERENCE

Jeffries, C. D., Holtman, D. F., and Guse, D. G. Rapid method for determining the activity of microorganisms on nucleic acids. *J. Bact.* **73:**590–591, 1957.

Method 3: Gram-Positive and Gram-Negative Organisms

Prepare DNase Test Agar with Methyl Green (Difco) according to manufacturer's directions. Slant in tubes or pour into plates.

1. Inoculate medium heavily in one spot; incubate overnight or longer at 35°C.
2. Read results as follows:

Positive: Colorless zone around growth.
Negative: No change in color of medium.

REFERENCE

Smith, P. B., Hancock, G. A., and Rhoden, D. L. Improved medium for detecting deoxyribonuclease-producing bacteria. *Appl. Microbiol.* **18:**991–993, 1969.

Method 4: Gram-Positive and Gram-Negative Organisms

1. Inoculate the test organism into 2 ml of sodium acetate buffer (pH 5.2, 0.1 M) containing 2 mg of 5-bromo-4-chloro-3-indolyl thymidine-3-phosphate.
2. Incubate at 37°C. Observe at 4 hours and after overnight incubation.

Positive: blue-green color on the bacteria and throughout the solution.

REFERENCE

Wolf, P. L., Horwitz, J., Mandeville, R., Vazquez, J., and von der Muehll, E. A new and unique method for detecting bacterial deoxyribonuclease in the clinical laboratory. *Tech. Bull. Reg. Med. Tech.* **39:**83–86, 1969.

ESCULIN HYDROLYSIS

Bile-Esculin Method (for Streptococci)

Prepare Bile-Esculin Agar (Difco) according to manufacturer's directions, but omit the horse serum. Prepare as slants.

1. Inoculate bile-esculin agar by streaking the slant.
2. Incubate overnight at 35°C. (If slant is inoculated heavily, may become positive in a few hours.)

   Positive: blackening of agar

   Only group D streptococci and rare strains of group Q or viridans streptococci will grow on this medium and hydrolyze the esculin. Other streptococci will not grow, and esculin hydrolysis by these organisms cannot be determined on this medium.

REFERENCE

Facklam, R. R. Comparison of several laboratory media for presumptive identification of enterococci and group D streptococci. *Appl. Microbiol.* **26:**138–145, 1973.

GELATIN LIQUEFACTION

Method 1: Stab Method

Prepare 12% gelatin in nutrient broth. Dispense into tubes as deeps. Autoclave at 121°C for 12 minutes.

1. Inoculate gelatin deeps by stabbing. Incubate at 20–22°C for 30 days.
2. To detect liquefaction, place tubes in refrigerator for 30 minutes. Remove and observe for liquefaction. Continue to incubate until liquefaction occurs or until the 30-day period is over.

   Strong positive: liquefaction within 3 days.
   Weak positive: liquefaction after 3 days.

REFERENCES

Edwards, P. R. and Ewing, W. H. *Identification of Enterobacteriaceae,* 3rd ed. Burgess Publishing Co., Minneapolis, 1972, p. 345.

Lennette, E. H., Spaulding, E. H., and Truant, J. P. *Manual of Clinical Microbiology*, 2nd ed. American Society for Microbiology, Washington, D.C., 1974, p. 912.

Method 2

Prepare 12% gelatin in nutrient broth. Autoclave at 121°C for 12 minutes; pour into plates.

1.   Inoculate plate in one spot. Incubate at 30°C for 3 days.
2.   Flood plate with mercuric chloride solution ($HgCl_2$, 12 g; distilled water, 80 ml; conc. HCl, 16 ml).
3.   Read results as follows:

     Positive: Clear zone around growth.

REFERENCES

Frazier, W. C. A method for the detection of changes in gelatin due to bacteria. *J. Infect. Dis.* **39**:302–309, 1926.

Cowan, S. T. and Steel, K. J. *Manual for the Identification of Medical Bacteria*. Cambridge Press, London, 1970, p. 156.

Rapid Method

1.   Inoculate 0.5 ml of saline with a heavy loopful of growth.
2.   Insert a strip of exposed, undeveloped x-ray paper (approximately 1 × 1¼ in.) into the saline suspension.
3.   Incubate in a heating block or water bath at 37°C. Observe 1, 2, 3, 4, 24 hours for removal of the green gelatin emulsion from the strip.

     Positive: Appearance of the transparent blue strip support.
     Negative: Strip remains green.

REFERENCE

Blazevic, D. J., Koepcke, M. H., and Matsen, J. M. Incidence and identification of *Pseudomonas fluorescens* and *Pseudomonas putida* in the clinical laboratory. *Appl. Microbiol.* **25**:107–110, 1973.

GLUCONATE OXIDATION

1.   Add a gluconate substrate tablet (Key Scientific Products Co., Los Angeles) to 1 ml of distilled water.

2. Inoculate with a heavy loopful of organism. Incubate overnight at 35–37°C.

3. Add a Clinitest tablet (Ames Co., Elkhart, Ind.)

4. Read the results as follows:

    Positive: green-yellow-orange precipitate.
    Negative: blue.

REFERENCES

Haynes, W. C. *Pseudomonas aeruginosa*—Its characterization and identification. *J. Gen. Microbiol.* **5:**939–950, 1951.

Gaby, W. L., and Free, E. Differential diagnosis of pseudomonas-like microorganisms in the clinical laboratory. *J. Bact.* **76:**442–444, 1958.

HIPPURATE HYDROLYSIS

Prepare 1% sodium hippurate in heart infusion broth. Dispense in 5-ml aliquots into screw cap tubes; tighten caps after sterilization.

1. Inoculate hippurate medium with 1–2 drops of an overnight broth culture of streptococci. Inoculate a positive control, group B streptococcus, and a negative control, group A streptococcus, in the same manner.

2. Incubate tubes for 42–66 hours at 35°C.

3. Centrifuge and remove 0.8 ml of clear supernate into a Kahn tube.

4. Add 0.4 ml of ferric chloride (12 g $FeCl_3 \cdot 6H_2O$ dissolved in 100 ml of dilute (5.4 ml conc. HCl to 94.6 ml dist. H_2O) HCl.

5. Shake mixture and let stand with occasional shaking for 10 minutes.

    Positive: a heavy, cloudy precipitate persisting after 10 minutes.
    Negative: clearing of the initial precipitate within 10 minutes.

REFERENCE

Facklam, R. R., Padula, J. F., Thacker, L. G., Wortham, E. C., and Sconyers, B. J. Presumptive identification of group A, B, and D streptococci. *Appl. Microbiol.* **27:**107–113, 1974.

HIPPURATE HYDROLYSIS (RAPID)

Prepare a 1% aqueous solution of sodium hippurate and dispense in 0.4 ml aliquots. Cork and store at −20°C.

1. Thaw tubes of sodium hippurate substrate.
2. Emulsify several small colonies or one large colony of beta hemolytic streptococci in a tube of substrate. The suspension should be very cloudy.
3. Inoculate positive and negative control organisms, group B and group A streptococci, respectively.
4. Incubate tubes in a heating block at 37°C for 2 hours.
5. Add approximately 0.2 ml (5 drops) of ninhydrin reagent (3.5 g ninhydrin in 100 ml of a 1 : 1 mixture of acetone and butanol) to each tube. DO NOT SHAKE TUBE.
6. Continue incubation for 10 minutes. DO NOT INCUBATE LONGER THAN ½ HOUR: FALSE POSITIVES COULD OCCUR.
7. Remove tubes and immediately record results.

Positive: deep purple.

Negative: no change or a very faint purple.

REFERENCE

Hwang, M. and Ederer, G. M. Rapid hippurate hydrolysis method for presumptive identification of group B streptococci. *J. Clin. Microbiol.* **1:**114–115, 1975.

INDOLE TEST

Kovacs' Method

1. Inoculate a tryptophane broth or other appropriate medium (i.e., MIO, SIM).
2. After 24–48 hours incubation at 35°C, add about 0.5 ml of Kovacs' reagent.

Positive: red color in the surface layer.

Kovacs' reagent: Dissolve 10 g of p-dimethylaminobenzaldehyde in 150 ml of amyl, isoamyl or butyl alcohol. Heat in a 56°C water bath until dissolved. Cool. Slowly add 50 ml of conc. HCl. Store in a glass-stoppered brown bottle in the refrigerator. This reagent should be a light yellow in color.

REFERENCES

Kovacs, N. Eine vereinfachte Methode zum Nachwein der Indolbildung durch Bacterien. *Z. Immunitatsforsch. Exp. Ther.* **55:**311–315, 1928.

Gadebusch, H. H., and Gabriel, S. Modified stable Kovacs' reagent for the detection of indol. *Am. J. Clin. Pathol.* **26**:1373–1375, 1956.

Ehrlich's Method

1. Inoculate a tryptophane broth or other medium containing tryptophane (e.g., 2% tryptone, peptone broth).
2. After 48 hours incubation at 35°C add 1 ml of xylene. Shake vigorously; let stand for 1–2 minutes. Slowly add about 0.5 ml of Ehrlich's reagent down the side of the tube. Do not shake.

   Positive: red ring just below the xylene layer.
   Ehrlich's reagent: Dissolve 1 g of *p*-dimethylaminobenzaldehyde in 95 ml of 95% ethyl alcohol. Slowly add 20 ml of conc. HCl. Store in a glass-stoppered brown bottle in the refrigerator. The reagent should be a light yellow color.

REFERENCES

Böhme, A. Die Anwendung der Ehrlichselen Indolreaktion für bakteriologische Zwecke. *Zentralbl. Bakt. Parasit. Abt. 1. Jena* **40**:129–133, 1906.
Lennette, E. H., Spaulding, E. H. and Truant, J. P. *Manual of Clinical Microbiology,* 2nd ed. American Society for Microbiology, Washington, D.C., 1974, p. 932.

Rapid Method

1. Inoculate 0.3 ml of 1% tryptone broth with a heavy loopful of growth from a solid medium (do not use from MacConkey's or EMB).
2. Incubate in a heating block or water bath at 37°C for 4 hours.
3. Add a few drops of Kovacs' reagent.

   Positive: red color in the surface layer.

REFERENCE

Blazevic, D. J. *Laboratory Procedure in Diagnostic Microbiology.* Telstar Productions, Inc., St. Paul, Minn., 1974, p. 110.

Spot Test

1. Moisten filter paper in a Petri dish with indole-spot reagent. Do not soak the filter paper; just moisten it.
2. Using growth from a medium containing tryptophane (e.g., blood agar

plate), place a loopful of organism on the filter paper. (Do not use growth from MacConkey's or EMB.)

Positive: development of a red color within a few seconds.

Indole-spot reagent: Dissolve 5 g of *p*-dimethylaminobenzaldehyde in 100 ml of 10% HCl. Store in glass-stoppered brown bottle in the refrigerator. The reagent should be a light yellow color.

REFERENCE

Vracko, R., and Sherris, J. C. Indole-spot test in bacteriology. *Am. J. Clin. Pathol.* **39**:429–432, 1963.

LECITHINASE TEST

Prepare egg yolk agar as follows:

Peptone	20 g
Na_2HPO_4	2.5 g
NaCl	1.0 g
$MgSO_4$	0.05 g
Dextrose	1.0 g
Agar	12.5 g
Distilled water	500 ml

Adjust pH to 7.4. Autoclave at 121°C for 15 minutes. Cool to 50–55°C. Add antibiotic-free egg yolk; mix and pour into plates. (Scrub egg and soak in disinfectant for 1 hour before separating aseptically.) For use with anaerobic organisms the medium should be stored in an anaerobic environment if not used within 4 hours of preparation.

1. Streak or spot organism on egg yolk agar plate.
2. Incubate 48 hours at 35°C in appropriate atmosphere for organism being tested.
3. Read results as follows:

Positive: opaque zone around growth

REFERENCE

McClung, L. S., and Toabe, R. The egg yolk plate reaction for the presumptive diagnosis of *Clostridium sporogenes* and certain species of the gangrene and botulinum groups. *J. Bact.* **53**:139–147, 1947.

LYSINE DEAMINASE

Prepare Lysine Iron Agar (LIA) according to manufacturer's directions. Slant so there is at least a 1-in. butt.

1. Inoculate LIA by streaking the slant and stabbing the butt twice.
2. Incubate at 35°C overnight.

   Lysine deaminase positive: red slant.
   Lysine deaminase negative: purple slant.

REFERENCE

Edwards, P. R., and Fife, M. A. Lysine iron agar in the detection of *Arizona* cultures. *Appl. Microbiol.* **9**:478–480, 1961.

MALONATE UTILIZATION

Standard Method

Prepare modified malonate broth according to manufacturer's directions.

1. Inoculate the broth with a 3-mm loopful of a broth culture or from a 18–24 hour culture on solid medium.
2. Incubate at 35°C for 24–48 hours.

   Positive: alkaline reaction, indicator turns blue.
   Negative: no change, indicator remains green.

REFERENCE

Ewing, W. H., Davis, B. R., and Reavis, R. W. Phenylalanine and malonate media and their use in enteric bacteriology. *Publ. Health Lab.* **15**:153–167, 1957.

Rapid Method

Prepare modified malonate broth according to manufacturer's directions. Dispense in 0.5-ml aliquots.

1. Inoculate the broth with a heavy loopful of an overnight growth from triple sugar iron agar or sheep blood agar (do not use growth from MacConkey's, as false-negatives may occur from this medium).

2.   Incubate in a heating block at 37°C for 3 hours.

Positive: blue color.
Negative: green or yellow color.

REFERENCE

Blazevic, D. J. *Laboratory Procedures in Diagnostic Microbiology,* 2nd ed. Telstar Productions, Inc., St. Paul, Minn., 1974, p. 111.

METHYL RED TEST

Standard Method

1.   Inoculate MR-VP medium; incubate at 30°C for 5 days.
2.   Add 5 or 6 drops of methyl red reagent per 5 ml of culture.

Positive: bright red color.
Negative: yellow or orange.

Incubation for 48 hours at 35°C is satisfactory for the majority of the Enterobacteriaceae. If equivocal results are obtained at this time the test should be repeated on cultures incubated longer.

Methyl red reagent: Dissolve 0.1 g methyl red in 300 ml of 95% ethanol. Add distilled water to make up 500 ml.

REFERENCES

Clark, W. M. and Lubs, H. A. The differentiation of bacteria of the colon-aerogenes family by the use of indicators. *J. Infect. Dis.* **17**:161–173, 1915.

Edwards, P. R. and Ewing, W. H. *Identification of Enterobacteriaceae,* 3rd ed. Burgess Publishing Co., Minneapolis, 1972, p. 348.

Rapid Method

1.   Inoculate 0.5 ml of MR-VP broth in a 13 × 100 mm test tube with one colony. Incubate at 37°C for 18 hours.
2.   Add one drop of methyl red reagent.
3.   Read results as follows:

Positive: bright red color.
Negative: yellow or orange.

Methyl red reagent: Dissolve 0.5 g of methyl red in 300 ml of 95% ethanol. Add 200 ml of distilled water.

REFERENCE

Barry, A. L., Bernsohn, K. L., Adams, A. P., and Thrupp, L. D. Improved 18-hour methyl red test. *Appl. Microbiol.* **20:**866–870, 1970.

NITRATE REDUCTION

Standard Method

Prepare Nitrate Broth according to manufacturer's directions. Dispense into tubes with inverted Durham inserts. (Alternatively, prepare the medium as semisolid by incorporating 0.25% agar into it.)

1. Inoculate medium, and incubate for 24–48 hours at 35°C.
2. Mix equal parts of Nitrate Solution A and B just before testing, and add 0.1 ml of the mixture to the culture.

   Positive: red color within 1–2 minutes.

   If test is negative, add a very small pinch of zinc dust to the tube.

   Positive: absence of red color development.
   Negative: development of red color.

3. If bubbles are apparent in the Durham tube (or throughout the semisolid medium) the nitrate need not be added if the organism is a nonfermenter.

   Positive: bubbles.

   If the organism is a fermenter (e.g., *Enterobacteriaceae*), the gas could be hydrogen, so the nitrate reagents must still be added and the test interpreted as in item 2.

Nitrate Solution A: Dissolve 8 g of sulfanilic acid to 1000 ml of 5 N acetic acid (1 part glacial acetic acid to 2½ parts distilled water).

Nitrate Solution B: Add 6 ml of N,N-dimethyl-1-naphthylamine to 1000 ml of 5 N acetic acid. Store Solution B at room temperature.

Rapid Method

1. Inoculate 0.5 ml of Nitrate Broth with a heavy loopful of growth from an appropriate medium (e.g., TSI, blood agar, chocolate agar.)
2. Incubate in a heating block or water bath at 37°C.
3. Add one drop of Solution A and one drop of Solution B. Shake.

    Positive: red color.

    If test is negative, add a minute amount of zinc dust. Shake.

    Positive: absence of red color.

REFERENCES

Blazevic, D. J., Koepcke, M. H., and Matsen, J. M. Incidence and identification of *Pseudomonas fluorescens* and *Pseudomonas putida* in the clinical laboratory. *Appl. Microbiol.* **25**:107–110, 1973.

Schreckenberger, P. C. and Blazevic, D. J. Rapid methods for biochemical testing of anaerobic bacteria. *Appl. Microbiol.* **28**:759–762, 1974.

ONPG TEST

Prepare ONPG broth as follows:

Peptone water: Dissolve 1 g of peptone and 0.5 g of NaCl in 100 ml of distilled water. Autoclave at 121°C for 15 minutes.

ONPG solution: Dissolve 0.6 g of *O*-nitrophenyl-β-D-galactopyranoside in 100 ml of 0.01 *M* Na_2HPO_4. Sterilize by filtration; put in sterile bottle. Store at 4–10°C, protected from light.

ONPG broth: Aseptically add 25 ml of ONPG solution to 75 ml of peptone water. Aseptically dispense in 0.5-ml amounts into sterile 13 × 100 mm tubes. This broth is stable for one month at 4–10°C or longer if stored in the freezer. *Do not use if yellow.*

1. Inoculate 0.5 ml of ONPG broth with a heavy loopful of growth from TSI or KI.
2. Incubate in a heating block or water bath at 37°C for 1 or more hours.

    Positive: yellow color.
    Negative: colorless.

This test cannot be performed with yellow-pigmented organisms.

REFERENCE

Cowan, S. T., and Steel, K. J. *Manual for the Identification of Medical Bacteria.* Cambridge University Press, 1970, p. 122.

OPTOCHIN TEST

Paper disks containing 5 μg of ethylhydrocuprein hydrochloride, optochin, are commerically available.

1. Pick a suspect colony with a needle.
2. Streak the isolate in a single line over at least one half of a sheep blood agar plate.
3. Spread the inoculum with a sterile swab to provide a uniform bacterial lawn.
4. Place an optochin disk in the center of the inoculum; press the disk to the agar surface with a sterile forceps or inoculating needle.
5. Incubate aerobically (not in CO_2) for 18–24 hours at 35°C.
6. Examine the plate for inhibition of growth around the disk.

   Positive: A zone of inhibition of 18 mm or greater around the disk.
   Negative: No zone of inhibition or a zone <18 mm around the disk.

REFERENCE

Austrian, R. *Streptococcus pneumoniae.* In Lennette, E. H., Spaulding, E. H., and Truant, J. P., Eds. *Manual of Clinical Microbiology,* 2nd ed. American Society for Microbiology, Bethesda, Md., 1974, p. 112.

OXIDASE TEST

Kovacs' Method

(Recommended for Nonfermenters and Miscellaneous Gram-Negative Bacteria).

1. Place filter paper in a Petri dish; saturate with 0.5% tetramethyl-*p*-phenylenediamine HCl.
2. Using a *platinum* wire, pick a portion of the colony to be tested, and rub the colony on the filter paper.

   Positive: appearance of dark purple color *within 10 seconds.* With this technique, there should be enough of the colony left for subculturing.

Alternate Procedure

Place a few drops of the oxidase reagent directly on colonies on a plate. Colonies producing oxidase become pink and then purple. The reagent does not interfere with the Gram stain, so that purple colonies may be stained even though not viable. The oxidase reagent may be divided into small aliquots and stored at −20°C. Thaw one aliquot as needed, and do not use for more than one day. Check reagent daily with a known oxidase-positive organism. The Kovacs' oxidase test is more sensitive than the method described below.

REFERENCES

Kovacs, N. Identification of *Pseudomonas pyocyanea* by the oxidase reaction. *Nature* **178**:703, 1956.

Lennette, E. H., Spaulding, E. H. and Truant, J. P. *Manual of Clinical Microbiology*, 2nd ed. American Society for Microbiology, Washington, D.C., 1974, p. 271.

Indophenol Method

1. Inoculate a nutrient agar slant; incubate at 35°C for 18–24 hours, no longer. Also inoculate an agar slant with *Aeromonas* for a positive control.
2. Add 2–3 drops Reagent A and Reagent B to the growth on the slant. Tilt the tube so that reagents are mixed and flow over the growth.

    Positive: blue color in the growth within 2 minutes.

    Weak or doubtful reactions that occur after 2 minutes should be ignored.

    Reagent A: Dissolve 1 g of alpha napthol in 100 ml 95% ethanol.
    Reagent B: Dissolve 1 g of *p*-aminodimethylaniline HCl (or oxalate) in 100 ml of distilled water. Reagent B should be stored in the refrigerator and should be freshly prepared each month.

REFERENCE

Ewing, W. H. and Johnson, J. G. The differentiation of *Aeromonas* and C27 cultures from *Enterobacteriaceae*. *Int. Bull. Bact. Nomencl. Toxon.* **10**:223–230, 1960.

PHENYLALANINE DEAMINASE

Method 1

Prepare phenylalanine agar in long slants according to manufacturer's directions.

1. Inoculate the phenylalanine agar slant by streaking heavily. Incubate at 35°C for 4 hours or 18–24 hours.
2. Allow 4–5 drops of 10% (W/V) ferric chloride to run over the growth.

Positive: dark green color in the fluid and slant surface.

REFERENCE

Ewing, W. H., Davis, B. R., and Reavis, R. W. Phenylalanine and malonate media and their use in enteric bacteriology. *Publ. Health Lab.* **15:**153–167, 1957.

Method 2

Prepare Phenylalanine-Urea (PU) medium as follows:

Yeast extract	1.0 g
$(NH_4)_2SO_4$	2.0 g
NaCl	3.0 g
K_2HPO_4	1.2 g
KH_2PO_4	0.8 g
DL-Phenylalanine	5.0 g
(or L-phenylalanine)	(2.5 g)
Distilled water	975 ml

Dissolve all ingredients; autoclave at 121°C for 15 minutes. Aseptically add 25 ml of Urea Agar Concentrate (Difco). Mix well, and aseptically dispense 0.5 ml into 13 × 100 mm sterile tubes. If medium is to be kept for long period of time, store at −20°C.

1. Inoculate 0.5 ml of PU medium with a colony of organism to be tested. Incubate overnight at 35°C.
2. Read the urease reaction as follows:

Positive urease: definite pink color.
Negative urease: yellow or faint tinge of pink.

3. Add 1–2 drops of 1% (V/V) HCl to adjust the medium to an acid pH (yellow).
4. Add 2 drops of 10% (W/V) ferric chloride.

Positive phenylalanine deaminase: dark green color.
Negative phenylalanine deaminase: yellow.

The phenylalanine deaminase reaction must be read within 10 seconds after adding the ferric chloride, as the green color fades rapidly.

With this method *Proteus*, most *Klebsiella*, some *Enterobacter*, and *Yersinia* will be positive, for urease. All other Enterobacteriaceae will be negative.

REFERENCE

Ederer, G. M., Chu, J. H., and Blazevic, D. J. Rapid test for urease and phenylalanine deaminase production. *Appl. Microbiol.* **21**:545, 1971.

Rapid Method

1. Inoculate 0.5 ml of PU medium with a large loopful of growth from TSI or other solid medium.
2. Incubate in a heating block or water bath at 37°C for 2 hours.
3. Read urease and phenylalanine deaminase reactions as under Method 2.

PHOSPHATASE

Method 1

Prepare nutrient agar according to manufacturer's directions; autoclave at 121°C for 15 minutes; cool to 50–55°C. For each 100 ml of medium add 2 ml of 0.5% sodium phenolphthalein diphosphate (sterilized by filtration). Pour into plates.

1. Inoculate phenolphthalein agar plates.
2. Incubate overnight at 35°C.
3. Hold the plate over an open bottle of ammonia (carry out procedure in a fume hood).

   Positive: colonies turn pink.

REFERENCE

Barber, M. and Kuper, S. W. A. Identification of *Staphylococcus pyogenes* by the phosphatase reaction. *J. Pathol. Bact.* **63**:65–68, 1951.

Method 2

1. Inoculate a heavy loopful of growth into 0.3 ml of p-nitrophenylphosphate solution (0.01 M p-nitrophenylphosphate in 0.1 M citrate buffer, pH 4.8).
2. Incubate up to 6 hours at 37°C.
3. Add 0.3 ml of 0.04 M glycine-NaOH buffer (pH 10.5).

   Positive: development of yellow color.

REFERENCE

Kersters, K. and DeLey, J. Enzymic tests with resting cells and cell-free extracts. In Norris, J. R. and Ribbons, D. W., Eds. *Methods in Microbiology,* Vol. 6A. Academic Press, New York, 1971, Chapter 11.

Method 3

1. Inoculate a heavy loopful of growth into a tube containing 2.75 ml of 0.1 *M* sodium acetate buffer (pH 5.2) and 0.25 ml of 5-bromo-4-chloro-3-indolyl phosphate substrate (2 mg/0.25 ml of NN-dimethyl formamide).
2. Incubate at 37°C for 30 minutes to 4 hours.

   Positive: blue-green precipitate on bacteria (eventually the entire solution turns blue).

REFERENCE

Von der Muehll, E., Ludwick, M., and Wolf, P. L. A new test for pathogenicity of staphylococci. *Lab. Med.* 3:26–27, 1972.

STARCH HYDROLYSIS

Prepare a broth or agar (appropriate to the type of organism being tested) with 0.2% soluble starch added. Prepare either slants or plates from the agar.

1. Inoculate starch medium. Incubate in an appropriate atmosphere at 35°C overnight or until sufficient growth has occurred.
2. To a broth add a few drops of Gram's iodine. Read immediately.

   Positive: no change in color.
   Negative: blue color (may fade after a while).

3. For plates or slants, flood with Gram's iodine.

   Positive: medium is blue with colorless area around growth.
   Negative: medium is blue even around growth.

REFERENCE

Allen, P. W. A simple method for the classification of bacteria as to diastase production. *J. Bact.* 3:15–17, 1918.

UREASE

There are many different tests for urease production. These tests vary considerably in sensitivity, and should be selected for the desired purpose. Only three tests are listed here.

Christensen's Urea

Add 3 g of agar to 180 ml of water. Autoclave at 121°C for 15 minutes. Cool; aseptically add 20 ml of Urea Agar Concentrate (Difco). Mix thoroughly, and aseptically dispense in 5–6-ml amounts into sterile 16 × 125 mm tubes. Slant so there is a butt of at least 1 in.

1. Inoculate Christensen's Urea medium by streaking the slant.
2. Incubate at 35°C or appropriate temperature for organism being tested for up to 4 days.

   Positive: bright pink color on slant or through entire medium (extent of color indicates rate of urea hydrolysis).

   With this method, many members of the Enterobacteriaceae (*Proteus, Klebsiella, Enterobacter, Citrobacter*) will give positive reactions, as well as *Cryptococcus, Yersinia,* and some other organisms.

REFERENCE

Christensen, W. B. Urea decomposition as a means of differentiating *Proteus* and paracolon cultures from each other and from *Salmonella* and *Shigella* types. *J. Bact.* **52**:461–466, 1946.

Rapid Method (Modified Rustigian and Stuart)

Prepare urease test broth (Rustigian and Stuart) according to manufacturer's directions. Dispense in 3 ml amounts into sterile tubes.

1. Inoculate urea medium very heavily (3 loopsful from an agar culture). Mix well.
2. Incubate in a heating block or water bath at 37°C. Observe after 10 minutes, 60 minutes, and 2 hours.

   Positive: deep pink color.

   With this method, the only member of the Enterobacteriaceae that should be positive is *Proteus*.

REFERENCES

Rustigian, R. and Stuart, C. A. Decomposition of urea by *Proteus*. *Proc. Soc. Exp. Biol. Med.* **47**:108–112, 1941.

Edwards, P. R. and Ewing, W. H. *Identification of Enterobacteriaceae*, 3rd ed. Burgess Publishing Co., Minneapolis, 1972, p. 353.

Combination method: See Phenylalanine deaminase.

VOGES-PROSKAUER TEST

Barritt's Method

1. Inoculate MR-VP medium; incubate at 35°C for 48 hours.
2. Remove 1 ml of the culture and transfer to a clean test tube.
3. Add 0.6 ml of 5% α-naphthol in absolute ethanol; mix well.
4. Add 0.2 ml of 40% KOH (containing 0.5% creatine, if desired); shake well.

    Positive: red color within 5 minutes.

REFERENCE

Barritt, M. M. The intensification of the Voges-Proskauer reaction by the addition of α-naphthol. *J. Pathol. Bact.* **42**:441–453, 1936.

O'Meara's Method

Follow the procedure above, except add 1 ml of 40% KOH (containing 0.3% creatine) to 1 ml of the MR-VP culture; shake well. Final readings should be taken after 4 hours.

   Positive: red-pink color.

If results are equivocal, repeat test and incubate at 25°C for 48 hours.

REFERENCE

O'Meara, R. A. Q. A simple delicate and rapid method of detecting the formation of acetylmethylcarbinol by bacteria fermenting carbohydrate. *J. Pathol. Bact.* **34**:401–406, 1931.

Rapid Method

1. Inoculate one colony or a loopful of growth (from TSI, KI, agar slant, blood agar, chocolate agar or MacConkeys, *not* from EMB) into 0.2 ml of MR-VP broth.

2. Incubate in a heating block or water bath at 37°C for 4 hours.
3. Add 2–3 drops of each of the Barritt reagents (see above), shaking between
    addition of each reagent.

    Positive: cherry red color within 15 minutes.
    Negative: any other color.

REFERENCE

Barry, A. L. and Feeney, K. L. Two quick methods for Voges-Proskauer test. *Appl. Microbiol.* **15**:1138–1141, 1967.

INDEX